IET POWER AND ENERGY SERIES 49

Series Editors: Professor A.T. Johns
Professor D.F. Warne

T0258015

Electric Fuses

Other volumes in this series:

Electric Fuses
3rd Edition

A. Wright and P.G. Newbery

The Institution of Engineering and Technology

Published by The Institution of Engineering and Technology, London, United Kingdom

First published 1982 (0 90604 878 8)
Second edition 1995 (0 85296 825 6)
Third edition hardback 2004 (0 86341 379 X)
Third edition paperback 2004 (0 86341 399 4)
Reprinted with new cover 2012

The Institution of Engineering and Technology
Michael Faraday House
Six Hills Way, Stevenage
Herts, SG1 2AY, United Kingdom

www.theiet.org

British Library Cataloguing in Publication Data
Wright, Arthur, 1923–
   Electric fuses. – 3rd ed. – (IET power and energy series no. 49)
   1. Electric fuses
   I. Title II. Newbery, P.G. (P. Gordon) III. Institution of Electrical Engineers
   621.3'17

ISBN (10 digit) 0 86341 399 4
ISBN (13 digit) 978-0-86341-399-5

Typeset in India by Newgen Imaging Systems (P) Ltd, Chennai
First printed in the UK by MPG Books Ltd, Bodmin, Cornwall
Reprinted in the UK by Lightning Source UK Ltd, Milton Keynes

Contents

Preface

Fuses have been produced for over 100 years and there are now an extremely large number in use throughout the world. They perform the vital duty of protecting equipment and electrical networks and ensure that the effects of faults, which inevitably occur, are limited and that the continuity of supply to consumers is maintained at a high level.

Not only electrical engineers, but nearly all members of the public are aware of the existence of fuses and the general impression is that they are simple devices in all respects. While it is true that their construction is not complex, they must be designed and manufactured with great care to ensure that they will perform as required. Surprisingly, the arcing process which occurs when they are interrupting current is still not fully understood. Research is continuing on this topic with the object of producing fuses capable of meeting the ever-increasing performance demands made on them. In this connection the advent and rapid growth of semiconductor devices, with their limited overload capacities, has introduced particularly stringent requirements.

The authors consider the subject to be of great importance and interest and therefore decided to produce this book which attempts to outline the history and early development of fuses. It then describes recent theoretical investigations of the current-interruption process before describing the constructions of the wide range of fuses which are currently produced throughout the world. The later chapters deal with application requirements, the various national and international standards with which fuses must comply and, finally, the quality-assurance and inspection procedures which are adopted by manufacturers are described.

This Third Edition includes salient aspects that have arisen since the publication of the Second Edition in 1995, including:

- Overview on resettable fuses and fault limiters
- The further analysis of pre-arcing and arcing behaviour.
- Fuses for Telecommunication Power Supplier
- The retrofitting of expulsion fuses with automatic sectionalising links.
- Developments in chip fuses and automotive fuses.
- Application information on: benefits of fuses; IGBT protection; arc flash and power quality.
- An update on national and international standards.
- The inclusion of a Glossary of terms.

Acknowledgements

For this Third Edition, Gordon Newbery wishes to record his thanks to the following fuse experts, for their contributions:

Joan Bender
R. Douglas
D. Giblin
J. Gould
V. Granville
P.M. McEwan
T.W. Mennell
R. Mollett
P. Reinhold
P. Rosen
V. Saporita
T.J. Stack
R. Wilkins

He also wishes to thank the following companies for their help with the additional illustrations:

Cooper Bussmann
Schneider Electric, Public Distribution Products

Thanks are also extended to Cooper Bussmann, Cooper (UK) Limited, for facilities made available during the preparation of this Third Edition.

Finally, a special thank you to Avril Burton, for her inspiration and 'encouragement' for me to prepare this Third Edition and for her patience and dedication preparing the revisions to the manuscript including accommodating the numerous changes.

February 2004   P. G. Newbery

Obituary—Arthur Wright

Arthur Wright, was a leading authority on the protection of electrical power systems and especially current transformers and fuses. He passed away in 1996 aged 73. His lectures were enlightening, literally. At his inaugural lecture he had arcs leaping across high-voltage insulators, gigantic demonstration fuses operating and large electric motors braking, reversing and accelerating very rapidly. A lay preacher for many years, he even managed to incorporate electromagnetic demonstrations into his sermons.

In Arthur's early days at the University of Nottingham, there were no induction courses for new lecturers and 'mentors' with whom to discuss problems. Arthur filled this gap willingly, applying his common sense and good humour to inspire young academics. He provided practical help on how to improve lectures, guidance on research, along with important advice on how to buy a house and bring up young children.

Belying his quiet and unassuming manner, he was a dedicated person with resolute determination. He had the enviable ability to communicate and obtain the respect of people at all levels and his endeavours seldom, if ever, led to friction or bad feeling. His outstanding gift was an enthusiasm for life and people in general and his ability to relate to everybody. With Joan to whom he was happily married for 49 years, he shared a love of music and with their two children Barry and Sue he enjoyed watching his favourite football teams, Sunderland and Nottingham Forest.

Alan Howe and Gordon Newbery

List of principal symbols

A_a = cross-sectional area of arc column
A_e = cross-sectional area of electrode
C = capacitance
D = density of electrode material
d_a = diameter of arc column
E_a = energy input to column in δt seconds
E_J = ionisation energy of an atom of the element material
e_s = source EMF
i = instantaneous current
$K_{m,n}$, $K_{m+1,n}$ etc. = thermal conductivities of subvolumes m, n and $m + 1, n$ etc.
l = length of arc
l_a = length of arc column
L_c = circuit inductance
L_f = latent heat of fusion of element material
L_v = latent heat of vaporisation
m_t = total electrode mass which is melted
m_v = mass of electrode which is vaporised
N_a = number of atoms evaporated from the electrodes in δt seconds
N_d = number of atoms scattered out of arc in δt seconds
N_e = number of electrons scattered out of arc in δt seconds
N_g = number of atoms per gram of electrode material
n = number of notches in element
n_a = atomic density
n_e = electron density
R_a = resistance of arc column
R_{am} = element resistance at ambient temperature
R_c = resistance of circuit
R_f = resistance of fuse
t = time
t_a = arcing time
v = instantaneous voltage
v_a = voltage along arc column

V_{af} = voltage associated with anode fall

V_{cf} = voltage associated with the cathode fall

V_T = voltage associated with the thermal energy of the electrons which enter the anode

V_{wf} = voltage associated with the work function of the element material

V_J = vapour jet velocity

vol_a = volume of the arc column

X = ionisation fraction

α = resistance temperature coefficient of element

δt = time interval used for computations

Δh_c = heat energy conducted to a subvolume in time Δt

Δh_g = heat energy generated in a subvolume in time Δt

Δh_l = heat energy lost from a subvolume in time Δt

Δh_s = heat energy stored in a subvolume in time Δt

θ_a = temperature of arc column

$\theta_{m,n}, \theta_{m,n+1}$ etc. = temperatures above ambient at the centres of subvolumes m, n and $m, n+1$ etc.

λ = specific heat

σ = electrical conductivity of the column

Subscripts

1 = quantities at the beginning of a time interval δt

2 = quantities at the end of a time interval δt

Chapter 1

Introduction

Fuses are among the best known of electrical devices because most of us have quite large numbers of them in our homes and, unless we are extremely fortunate, we are made aware of their presence from time to time when one must be replaced because it has blown or, to use the official term, operated. They are basically simple and relatively cheap devices, although their behaviour is somewhat more complex than may be generally realised.

The underlying principle associated with fuses is that a relatively short piece of conducting material, with a cross-sectional area insufficient to carry currents quite as high as those which may be permitted to flow in the protected circuit, is sacrificed, when necessary, to prevent healthy parts of the circuit being damaged and to limit the damage to faulty sections or items to the lowest possible level. As an example, a fuse element a few centimetres long with a particular cross-sectional area could be used to protect an electrical machine winding containing a considerable length of conductor, maybe kilometres, of a cross-sectional area slightly greater than that of the fuse element. In this case the volume of conducting material to be sacrificed in the event of a fault would only be a tiny fraction of that being protected and the cost of the protection would clearly be acceptable.

Fuses incorporate one or more current-carrying elements, depending on their current ratings, and melting of these, followed by arcing, occurs when excessive overcurrents flow through them. They can be designed to interrupt safely the very highest fault currents that may be encountered in service, and, because of the rapidity of their operation in these circumstances, they limit the energy dissipated during fault periods. This enables the fuses to be of relatively small overall dimensions and may also lead to economies in the cost and size of the protected equipment.

Because of the above advantageous features, fuses have been and are used in a wide variety of applications, and it appears that the demand for them will continue at a high level in the future. They were undoubtedly incorporated in the earliest electric circuits in which the source power and value of the equipment were significant.

1.1 History of fuse development

Since the Second Edition, an earlier reference to fuses has come to light in 1774 by Edward Nairne. This was in the era of electrostatic electricity and Nairne was trying to safely discharge Leyden bottles (capacitors) using wires with length related to the stored energy, acting like resistors but when the discharge currents were too high, the wire acted like a fuse, disintegrating into small balls (unduloids).

An early reference to fuses occurred during the discussion following the presentation of a paper by A. C. Cockburn [1] to the Society of Telegraph Engineers in 1887 when W. H. Preece stated that platinum wires had been used as fuses to protect submarine cables since 1864, and Sir David Salomons referred to the use of fuses in 1874.

A considerable number of fuses must have been in use by 1879, and presumably the simple wire construction was not even then adequate for some applications because in that year Professor S. P. Thompson produced what he described as an improved form of fuse or cutout. It consisted of two iron wires connected together by a metallic ball, as shown in Figure 1.1. It was stated that the ball could be an alloy of lead and tin or some other conducting material of low melting point. When a sufficiently high current passed through the fuse for a long enough period, melting of the ball occurred and it fell, allowing the wires to swing apart and break the circuit. It should be remembered that most circuits in use up to about 1890 carried direct currents and in these circumstances the sudden separating action would undoubtedly be needed to achieve arc extinction.

A variation on Professor Thompson's design was patented in 1883 by C. V. Boys and H. H. Cunyngham. In their arrangement the current flowed through two leaf springs which were soldered together at their inner tips, as shown in Figure 1.2. Above a particular current the solder melted and allowed the strips to flex in opposite directions, thus giving a sudden break. Other physical arrangements based on this principle were produced, one being attributed to Sir W. Thomson.

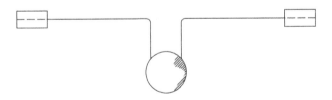

Figure 1.1   Fuse developed by Professor S. P. Thompson

Figure 1.2   Fuse patented by Boys and Cunyngham

Demonstrations of incandescent filament lamps had been given in Britain by J. Swan (later to become Sir Joseph Swan) in 1878, and almost simultaneously by T. A. Edison in the USA. Production of these lamps was started shortly afterwards and this caused a tremendous demand for electric lighting in public and private buildings. Initial installations included their own generating plants, but networks supplied from small central stations were soon in use.

Interesting detailed information about early installations is given in letters written by J. H. Holmes and Colonel R. E. Crompton to H. W. Clothier in 1932. Excerpts from these letters, which were included in Clothier's book entitled *Switchgear Stages* [2], are reproduced below. They clearly indicate that there is doubt about the identity of the person who first introduced fuses.

Letter from Mr J. H. Holmes:

Regarding the origin of fuses, I have always been uncertain as to who is entitled to the credit of being the first inventor, and am of the opinion that this is a very clear case of "Necessity is the Mother of Invention".

I have been looking up some records of what was known about fuses in the early "eighties", and in the first volume of "Electric Illumination", compiled by J. Dredge and published August, 1882, at the Offices of "Engineering", on page 630 it is stated that Edison's British Patent of April, 1881, appears to have been the first notification of lead safety wire. It also appears that Edison's device was called a "safety guard".

I think however that Swan used a device for the same purpose and before April, 1881, because "Cragside" near here, the seat of Sir W. G. (afterwards Lord) Armstrong was lighted with Swan lamps by the middle of December, 1880. Swan used tin foil for the fuse, and a strip of this was jammed between two brass blocks, so as to form part of the circuit, by a plug of wood and later of steatite, and I have samples of a combined switch and fuse, and a fuse only made in this way, and which were in use at Cragside. In a Swan United Electric Light Co's catalogue dated 1883, I find such fuses illustrated and called "safety-fusing bridges".

In the description of the Electric Lighting, on the Swan system, of the Savoy Theatre in "Engineering", March 3rd, 1882, "fusible safety shunts" are referred to as "not intended so much to guard against a danger which is next to impossible to occur in practical working, but to protect the lamps themselves from destruction from too powerful a current being transmitted through them". This seems to confirm what Campbell Swinton says about the Drawing Office at Elswick in 1882,* which you quote, and I note he also says that at the Paris Exhibition of 1881 there was "a vast array of switches, fuses, cut-outs, and other apparatus".†

In the same Commemoration Number, I give on page 471, an account‡ of my first experience of a really alarming "short", and as Mr. Raworth at once proposed to have a fusible wire, which was enclosed in a wooden pill box, put into the circuit, I have always thought of him as one of the inventors of the "fuse".

---

*"It is perhaps not generally known that fuses, as originally introduced by Swan, were designed not as a safeguard to protect the wires against overloading on short-circuits, but in order to

prevent the lamps from over-running. When I went to the Armstrong Works at Elswick in 1882, part of the drawing office had been electrically lighted by the Swan Company, and each incandescent lamp was fitted with a separate tinfoil fuse for this purpose. The precaution was, perhaps, a necessary one, as the lamps then cost 25s. each and were very fragile, while the arrangements for keeping a constant voltage were very crude". (Campbell Swinton, at the IEE Commemoration Meetings, February 1922, *IEE Journal*, 1922, Vol. 60, p. 494).

†"In 1881 the Electrical Exhibition was held in Paris, at which were gathered together, for the first time, a most comprehensive selection of all the wonderful electrical inventions of the preceding years . . . . There were shown in competition the then newly invented incandescent lamps of Swan, Edison, Lane Fox and Maxim, together with a vast array of switches, fuses, cut-outs, and other apparatus that had then just been designed to meet the requirements of the new method of both public and domestic illumination". (Campbell Swinton, IEE Commemoration Meetings, February 1922, *IEE Journal*, 1922, Vol. 60, p. 494).

‡The first steamship to carry electric arc lamps for interior illumination was the s.s. *City of Berlin*, and the s.s. *City of Richmond* was the earliest instance of the electric lighting of a vessel by the incandescent system, and was completed in June, 1881. Very soon after this date the s.s. *City of Rome* was fitted with a very complete installation by the late Mr. John S. Raworth, on behalf of Messrs. Siemens Bros. & Co. I joined his staff and assisted with the work, and was present at the first trials. I remember one rather alarming occurrence. We did not know much about fuses, and when we started up the plant I was horrified to see sulphurous smoke from the vulcanised rubber on the cables, rising from the top of the dado all around the music saloon, and turning the recently finished white-lead painting to a streaky black. I rushed off to the engine room to get the current turned off (we had no switches nearer) and, being thankful that we had not set the ship on fire, we bribed the painters to work all night to eradicate all evidence of the mishap". (Holmes, IEE Commemoration Meetings, February 1922, *IEE Journal*, 1922, Vol. 60, p. 471).

I have in my collection several examples of both tin foil and lead wire fuses in wooden boxes which date from the early "eighties", and these I hope to find room for in the showcase which is to contain my old electric lamps at the new Municipal Museum (Newcastle upon Tyne). I am very busy at present cataloguing and fixing the lamps on the sloping shelf which will carry them in the showcase.

Letter from Colonel R. E. Crompton:

In the year 1881 my firm got a definite order to light a country house, this was the house of Mr. Jesse Coope, a partner in the firm of Ind, Coope the brewers of Romford. Mr. Coope wanted his newly built house to be equipped throughout electrically and to my partner Mr. Harold Thomson,* the son of the great inventor and brother of Sir Courtauld Thomson, must be given credit for a very large share of the design of the fittings and the general arrangements, which commencing at Coope's house were carried out in a large number of country houses, and some town houses. So that early in the year 1882 we were entrusted by Shaw Lefevre, then First Commissioner of Works, to tender for the complete electrical installation of the Law Courts which were nearing completion.

The Law Courts installation was upwards of 2,000 Swan lamps and was fitted with every appliance that we had found necessary for domestic work and I am practically certain that all the branch circuits were then protected by fuses, and that the real inventor of fuses was Harold Thomson.

At Coope's house, Harold Thomson had a few bell hangers as his workmen. He then introduced lamps suspended from the ceiling by their own conductors, the whole system of controlling lights by switches fixed near the doorways, ceiling roses and last but not least the dividing up of each installation into a number of small circuits controlled by small switchboards.

During the years that followed, say from 1881 to 1885, Cromptons carried out a very large number of separate installations, to the best of my belief in all of these the branch circuits were protected by fuses. So that I feel practically certain that the fuses were used to protect the branch circuits in the various installations which were worked from the Central Station in Vienna † in 1886.‡

*"My first partner, Harold Thomson, was the inventor or designer of most of the common appliances that we now use, as nearly all the switches, plug contacts, fuse arrangements, the general arrangement of switchboards, the suspension of lamps by their own flexibles, were worked out by him and by Mr. Lundberg". (Crompton, IEE Commemoration Meetings, February 1922, *IEE Journal*, 1922, Vol. 60, p. 394).

†"The Vienna scheme was remarkable in that we had to supply 20 000 lamps in theatres 3/4 mile from the central station. We used a 5-wire system, with about 500 volts across the dynamo terminals, the distribution being by four large batteries of accumulators coupled between the five wires. The Willans engines were of 200 h.p. each and were electrically governed.... As our work at Vienna extended to those installations at the Grand Opera and other theatres we had to devise, for the first time, most of the electric appliances which are still in use in theatres. Monier and I worked out convenient methods of indicating the current in our feeders by measuring the fall of potential on a portion of our bus-bars, by using a delicate D'Arsonval galvanometer with magnetic control: in this way we were able to measure currents up to several thousand amperes within 1 per cent. During the progress of the Vienna work, which went on from 1884 to 1887, we attempted to commence house-to-house lighting in London, but found it very difficult to raise capital under the terms of the original Chamberlain Act". (Crompton, IEE Commemoration Meetings, February 1922, *IEE Journal*, 1922, Vol. 60, p. 394).

‡In another Letter, Col. Crompton recalls a bad short on the big switchboard at the Schenkenstrasse station, caused by one of his fitters, George Earthy (who afterwards became a teacher at the Battersea Polytechnic), and points out that this was "the first central station ever built in the world".

Several factors, including the concern for public safety, the cost and fragility of the lamps, referred to earlier, and the increasing level of the available volt-amperes under fault conditions, made evident the need for protective equipment, and, as a result, a number of workers sought to develop reliable fuses, there being no other obvious alternative protective devices at that time.

Figure 1.3   Fuse developed by Cockburn

A considerable amount of work was done to obtain an understanding of the processes involved during the melting of fuse elements. A particularly significant contribution was made by A. C. Cockburn [1], and details of his work are given in the paper referred to earlier. He was critical of the physical constructions and inconsistencies present in many of the fuses in use at the time and attempted to put the design of fuses on a sound engineering basis. He studied the effect of the heat conducted away from fuse elements, to their terminals and the connecting cables, and realised that this factor could significantly affect the minimum current at which a particular fuse would melt. He investigated the properties of conductors and took account of specific heat, thermal and electrical conductivity and other parameters in an attempt to select the materials which should prove most suitable for use as fuse elements. He recognised that materials which oxidise readily and significantly would be unsuitable because the fuse characteristics would change with time as a result. Having acquired a good understanding of fuse behaviour, he developed a fuse in which a weight was hung on the wire element as shown in Figure 1.3. In his design, unlike that of Professor Thompson, the current did not flow in the weight. He claimed that a 'magical result' was obtained, because the weight caused the wire to break when it became sufficiently molten and the performance obtained was more consistent than that of other fuses produced at the time. Tests which he did showed that fuses were not being applied in a consistent or scientific manner, there being instances where the minimum fusing currents were many times the rated currents of the circuits of equipment being protected. He suggested that a fuse should operate at 150–200 per cent of the rated current of the circuit being protected.

The fuses described above were usually mounted in wooden boxes, but the individual elements were not otherwise enclosed. As early as May 1880, however, T. A. Edison patented a fuse in which the wire element was enclosed in a glass envelope. This was not done to affect the fuse performance but to protect the surroundings from the effects of the disruption during operation. This is clear from the patent specification, which included the following statement: 'the small safety wire becomes heated and melts away, breaking the overloaded branch circuit. It is desirable, however, that the few drops of hot molten metal resulting therefrom should not be allowed to fall upon carpets or furniture, and also that the small safety-conductor should be relieved of all tensile strain; hence I enclose the safety-wire in a jacket or shell of nonconducting material'. Edison's devices did not contain filling material and credit for the filled cartridge fuse must undoubtedly go to W. M. Mordey who took out a patent in 1890. This patent described a fuselink with a fusible copper conductor,

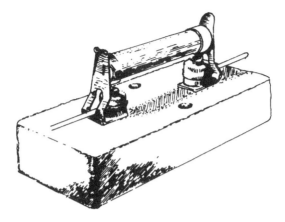

Figure 1.4   Cartridge fuse patented by Mordey

of either thin foil or one or more small-diameter wires, enclosed in a glass tube or similar vessel. It was stated that the tube should be wholly or partially filled with finely divided, semiconducting or badly conducting material which should prefer- ably be incombustible or non-flammable. Dry chalk, marble, bath brick, sand, mica, emery and asbestos were suggested as possible filling materials. The fuse produced by Mordey is illustrated in Figure 1.4.

Sand of controlled grain size and purity is today, still the preferred medium for arc extinction and heat transfer in high-performance fuses. In its arc extinction role the sand filler dissipates the energy of the arc by melting so that when cooled down it looks like a small lump of pumice stone. The ancient Romans, when they saw lightning strike sand in the desert, called the solid lumps of sand 'FULGURITE', and this is the term used by fuse engineers to describe the inside of the fuse once it has operated.

During the final decade of the 19th century, fuses represented the only form of protective equipment available and that they were produced in large ratings is evident from descriptions in H. W. Clothier's paper entitled "The construction of high-tension central-station switchgears, with a comparison of British and foreign methods" [3]. The relevant extract from the paper reads as follows:

The High-Tension Fuse most extensively used in Germany is not unlike the well-known Bates fuse, consisting of an open-ended tube of porcelain, ambroin, stabilit, or similar insulating material, with plug terminals at each end. The fuse-wire of copper or alloy is threaded through the tube and clamped by screws and plates or soldered to the terminals. For potentials of 2000 to 10000 volts the length of these tubes varies between 8 inches and 15 inches. Several fine wires are connected in parallel for the higher voltages, each wire being enclosed in a separate internal tube or otherwise partitioned by insulating material, so that each wire has a column of air to itself. Unlike the Bates fuse, there is no handle moulded with the tube, but flanges are provided at the ends, and moreover, in most cases, it is customary to have a long pair of wooden tongs close by the switchgear with which any fuse can be clutched while the operator is at a safe distance from it. Considering the massive tube fuses—from four to five feet long—used at Deptford and Willesden, and also expensive designs such as the oil-break fuses of home manufacture, it would appear that either we over-estimate the destructive effects caused in breaking high-tension circuits or else the necessity of blowing a fuse without destroying the fuse-holder is not considered a matter of importance in Germany. The variation in size and initial cost of the respective designs is very pronounced.

It is very common in German practice to construct high-tension switchboards without fuses or automatic devices of any kind in circuit with the alternators, and to have plug-fuses on the outgoing feeders, but no switches on these circuits. The fuses in such cases are placed in cellars or in basements away from the station attendants. There is thus no way of breaking a load on a feeder-circuit other than by blowing its fuse. The visible switchgear in these stations is only that required for the control of the machines under normal conditions. We were informed in one important station in Berlin that one man was never sent alone into the cellar or basement, but, for safety, was always accompanied by another to look after him.

A very popular fuse in the early 1900s was the 'Zed' fuse. This was produced by Siemens Brothers and by 1912 the total sales in Great Britain were already over a quarter of a million fuses. An illustration of the 'Zed' fuse, taken from an early Siemens Brothers catalogue in the UK, is shown in Figure 1.5.

These were fitted into distribution boards, often with a pleasing aesthetic appearance, an example of which is shown in Figure 1.6. The basic 'Zed' fuse concept is still in use today, see Section 4.3.2.

During the twentieth century, relay-based protective schemes have been produced, and these together with the circuit breakers they control are now used in conjunction with all the major items of plant in the generation and transmission networks. Nevertheless, because of the enormous growth in the amount of electrical equipment being used throughout the world there has been a steady increase in the demand for fuses and

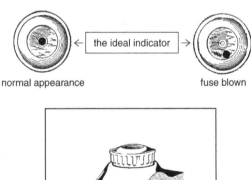

normal appearance   fuse blown

Figure 1.5   The complete 'Zed'

Figure 1.6   'Zed' distribution boards

this has caused investigators to study basic phenomena, such as the arcing process, and also to find ways of providing fuses with the characteristics needed to protect circuits and devices such as semiconductors and rotating machines. An examination of the literature shows that great progress has been made but even now some aspects of behaviour are still not fully understood and consequently research is continuing.

1.2   Basic requirements

In 1882, an *Electric Lighting Act* was passed by the British Parliament and this was amended six years later to form what was known as the *Electric Lighting Acts 1882 and 1888*. These acts specified conditions with which British electricity-supply companies had to comply and, interestingly, legal processes were outlined whereby gas-supply undertakings could seek to be relieved of their obligation to supply gas to areas in which their operations had become unprofitable because of the newly provided electric lighting. An important feature of the acts was that they gave the UK Board of Trade the responsibility of introducing regulations to secure the safety of the public and ensuring a proper and sufficient supply of electrical energy. As a result, several documents specifying regulations were produced by the Board of Trade in draft and then final forms, in the period up to 1910. Each of them included a clause stating that a suitable safety fuse or other circuit breaker must be inserted in each service line within a consumer's premises as close as possible to the point of entry, and contained within a suitable locked or sealed receptacle of fireproof construction, except in those cases where the service lines were protected by fuses in a street box.

By 1900, there was a proliferation of fuses of various shapes and sizes and the first reference to standardisation of fuses which the authors have found in the UK is an IEE paper by K. W. Kefford [4] in 1910. Three interesting extracts from his paper are as follows:

So long ago as 1905 an editorial in one of our electrical journals drew attention to the importance of standardising low-tension fuses, and, whilst regretting that we lagged far behind other countries in giving this latter the attention it deserved, took comfort in the assurance that, at some future date, we should fall into line

Makers of fuses now display in their catalogues so many types and so many designs of each type that it is difficult to escape the conclusion that the fuses are ornamental adjuncts to a switchboard and may be chosen in a style most suitable to, say, the scrollwork which supports the clock.

The complete specification of a line of fuses will embrace:

(a) A definition of the "marked" or "rated" current in terms of the "limiting" current (also called the "normal fusing" current).
(b) A standard range of rated-current values and voltages.
(c) A definition of one or more points on the "time-overload" curve of each fuse.
(d) Regulations as to non-interchangeability, temperature rise, freedom from deterioration, and perfect operation under all conditions.
(e) Specifications for the standard method of carrying out short circuit, temperature rise, and overload tests.

In framing suggestions with regard to the above points, due consideration is demanded for the user, the workman, the retailer, and the manufacturer; for example, the quantitative properties of the fuses should lend themselves to simple control in the factory, the number of types should be reduced to a minimum, the price should be moderate, etc.

There were however no British specifications that defined the performances required of fuses until *BS 88* was introduced in 1919. This standard was limited to fuses for rated currents up to 100 A at voltages not exceeding 250 V. It included definitions of terms such as 'fuse carrier' and 'fusing current' and specified the maximum short-circuit currents that fuses rated at 10, 30, 60 and 100 A should be able to interrupt and also the corresponding fusing currents. In addition, there were clauses specifying dimensions and contact materials but, somewhat surprisingly, operating time limits were not included.

The IEE 'Regulations for the electrical equipment of buildings' (8th edition), issued in 1924, included a clause requiring that a fuse carrying a 100 per cent over-current should operate in 1 min if the wire was of tinned copper or 2 min if it was of lead–tin alloy.

The revision of *BS 88*, issued in 1931, included a requirement that a fuselink should blow in less than 30 min at 1·9 times the rated carrying current and should be able to withstand 1·6 times the rated current for at least 30 min. Operating-temperature limits and short-circuit capability were also specified.

Since that time there have been further revisions of *BS 88* and a number of other standards have been produced in Britain. There are also corresponding standards in the United States and European countries. In an attempt to clarify the situation the International Electrotechnical Commission has also produced standards to ensure that:

(*a*)   each fuse is so designed and produced that it will, throughout its life, allow the circuit in which it is included to operate continuously at currents up to the rated value

(*b*)   each fuse will operate in a sufficiently short time when any current above a certain level flows, because of an overload, to prevent damage to the equipment being protected

(*c*)   in the event of a fault developing on a network or piece of equipment, fuses will operate to limit the damage to a minimum and confine it to the faulted item.

(*b*) and (*c*) imply that fuses must possess inverse time/current characteristics and for any applications these should be so selected that discrimination is achieved during fault conditions and unnecessary disconnection is thereby avoided.

The requirements specified in the various national and international standards to which fuses are produced cover both technical performance and dimensions. They are dealt with in some detail in Chapter 8.

1.3   Fuse types and constructions

At this point the treatment will be general and introductory in nature to enable readers, who are not familiar with fuse terminology and the wide range of fuses available, to acquire background information.

1.3.1   Classifications

Fuses are classified into three categories: high-voltage (HV), low-voltage (LV) and miniature. These categories are recognised nationally by the British Standards Institution and internationally by the International Electrotechnical Commission. The division between HV and LV occurs at 1000 V AC, and the miniature designation is clearly associated with physical dimensions.

1.3.2   Basic constructions

Fuses are produced in a number of constructional forms and in general can be classified into 'enclosed' or 'semi-enclosed'.

1.3.2.1   Enclosed fuses

As the name implies enclosed fuses are designed to keep the arc products within the enclosure during specified operation. Enclosed fuselinks are often of cylindrical form and are referred to as cartridge fuses.

Cartridge fuses, which form a very important and certainly the most numerous group, may be designed for HV, LV or miniature applications. For LV applications, the fuselink, which is replaceable, is often fitted into a fuse holder that consists of a fuse

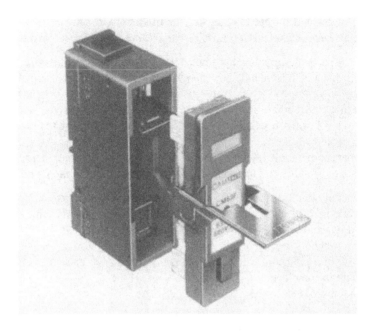

Figure 1.7   Low-voltage cartridge fuse

carrier and a fuse base, a particular example being illustrated in Figure 1.7. Cartridge fuselinks for low current ratings contain a single element while those for higher ratings are fitted with a number of parallel-connected elements. It is essential that elements should have a high resistance to oxidation so that significant changes or deterioration do not occur during fuse life. In the past this was achieved by making them almost exclusively of either silver or plated copper. To reduce costs, however, plain copper elements have been introduced and they are now widely used, particularly in low-voltage industrial fuses. The effects of oxidation have been reduced to acceptable levels by so designing the elements that they operate at relatively low temperatures at their rated currents.

Wire elements are used in fuses with low current ratings but it is necessary to employ strips with one or more portions of reduced cross-sectional area along their lengths in fuses with ratings of more than about 10 A. The elements are attached to plated copper or brass end caps which, together with the body, form a complete enclosure or cartridge. For high-breaking-capacity fuselinks the space within the enclosure is usually filled with quartz of controlled grain size and high chemical purity. Fuselink bodies must be good electrical insulators, and, in addition, must be mechanically robust and capable of withstanding thermal shock during operation. Ceramic and glass were used almost exclusively for this purpose in the past, but, more recently, glass reinforced plastics have been introduced and have gained a measure of popularity. Electrical connection can be made to a fuselink by spring clip contact on to the end caps – these are called ferrule-type fuselinks and they are usually restricted to applications with current ratings up to 100 A. Alternatively, electrical connection

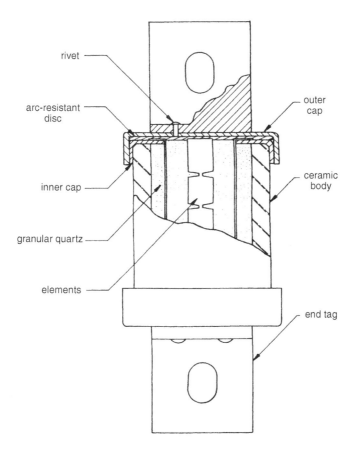

Figure 1.8   Cross-sectional view of a cartridge fuselink

can be made by integral spade-type terminations that have holes through which bolts may be fitted. Figure 1.8 shows a cross-sectional view through a typical cartridge fuselink.

There are other shapes for enclosed fuses and two popular types which are made in considerable volume, are:

- 'Chip fuses' for use on printed circuit boards with surface mounting (see Section 6.1.3.1).
- 'Blade-type fuses' for use in automotive applications (see Section 6.3).

Enclosed fuselinks usually have fuse elements, which are surrounded by air or with sand. Some fuses have other mediums surrounding the fuse elements, including liquid fuses (see Section 5.1.2).

Liquid fuses are in the HV category. A fuse of this type has a liquid-filled glass tubular body and within this is a short element of wire or notched strip, held in tension by a spring. When the element melts, a rapid separation is produced causing the resulting arc to be extended and extinguished in the liquid filling. The current-breaking capacities of these fuses are rather limited and although many are in use and giving good service, they are no longer recommended for new installations.

Vacuum fuses have also been produced for medium voltage applications and used with associated switchgear.

1.3.2.2   Semi-enclosed fuses

The most popular form of semi-enclosed fuses is the expulsion fuse (see Section 5.1.1).

The expulsion fuse, which belongs to the HV category, contains a mechanism to move the fuselink away from one of its contacts when the element melts and as a result, a long gap is introduced in air in series with the fuse. The arcs in the air and within the fuse are extinguished by the expulsion effects of the gases produced by the arc, provided that the current level is not too high. In practice, this type of fuse is only able to interrupt currents up to about 8000 A.

'Rewirable' semi-enclosed fuses can still be found in the LV category (e.g. Figure 1.9).

A fuse of this type consists of a fuse-holder made up of a fuse base and a fuse carrier, the latter containing the element, which is invariably in wire form. Elements are directly replaceable and hence these fuses are often referred to as rewirable. They are of low breaking capacity, being typically capable of interrupting currents up to only 2000 A. In the past, these fuses were produced and installed in large numbers in the UK and many of them are still in service. The demand for them has declined in recent years however, because their performance is inferior to that of cartridge fuses.

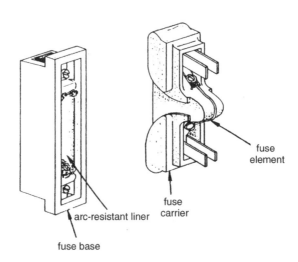

Figure 1.9   Semi-enclosed fuse

1.3.3 'Resettable fuses' and fault limiters

Readers may come across products referred to as 'resettable fuses', 'permanent power fuses' and even 'superconducting fuses'. These products are essentially 'fault limiters' where their resistance increases by typically two to five orders of magnitude when called upon to react. They return again to a low-resistance state after the event.

Fuses, however, by definition (see Glossary of terms) are devices whose current-carrying element(s) melts and permanently breaks the circuit. Fuses are thus 'one-time' devices and fault limiters of the above kind are therefore *not* fuses.

For completeness, an overview of 'Resettable fuses' is provided in this section. For further reading, a technical review of such current limiters was undertaken in 1995 by Lindmayer and Suchbert [5].

Fuses that can be 'reset' have the attraction of easing the logistical supply of replacements and simplification of the restoration of supply. However, this is offset by the need to evaluate the potential safety hazard for applications including the performance capability for all over-current conditions.

One successful area where 'resettable fuses' have been extensively used is for electronic applications with ratings up to around

- 600 V, trip current 0.35 A;
- 30 V, trip current 7.5 A.

The latter being used in automotive applications.

Such 'resettable fuses' are usually positive temperature coefficient (PTC) devices. A PTC device is a non-linear thermistor that limits its current. PTC devices are either ceramic, CPTC, or polymer, PPTC. Because under a fault condition all PTC devices go into a high-resistance state, normal operation can still result in hazardous voltage being present in parts of the circuit. It is important that the circuit designers recognise critical differences between the two devices. Fuses are current interruption devices and once a fuse operates, the electrical circuit is broken and there is no longer current flowing through the fuse. This electrical interruption (or open circuit) is a permanent condition. However, once a PTC device trips, there is a small amount of current flowing through the device. PTC devices require a low joule heating leakage current or external heat source in order to maintain their tripped condition. Once the fault condition is removed, this heat source is eliminated. The device can then return to a low-resistance status and the circuit is restored.

The technology has been commercially extended [6] to LV applications up to 63 A, 690 V and uses fault limiters in conjunction with circuit breakers.

Another technology commercially exploited was the 'permanent power fuse' or the 'sodium' fuse [7]. Here the fuse element was sodium and used for back-up protection in motor applications. However, cost and degradation of technical properties with repeated operation has led to its decline.

Regarding superconducting current limiters, Lindmayer and Schubert [5] indicated problems associated with the transition between superconducting and normal conducting state. Subsequently a prototype has been produced but it may well be another decade before commercial exploitation can be fulfilled based on progress being made in other areas for high-temperature superconductors.

1.3.4   The antifuse

Readers may also come across reference to such a device in CMOS technology. An antifuse is the opposite of a regular fuse. An antifuse is normally open circuit until a programming current, typically 5 mA, is forced through it.

1.4   World production

The world's usage of electricity has increased steadily over the years and is still doing so, the current usage being more than 10^7 GW h per annum. This is obtained from an installed capacity of about 3500 GW.

Fuselinks are required for replacement purposes and in new installations. Projections based upon UK production, and related to the above usage figures for electricity and installed capacity, give an estimated production of over ten thousand million, 10^{10}, fuselinks per annum. There are, however, an extremely large number of different types and current ratings and therefore only a small number of types have sufficient volume to warrant fully automated manufacture. The majority of the fuselinks produced are of the miniature and LV domestic types and, although the number of LV industrial fuselinks is small by comparison, it is estimated that several hundred million are produced each year.

HV fuselinks, because of their more limited applications, are produced in relatively small quantities but they nevertheless play a vital role in protecting power systems.

Chapter 2

Pre-arcing behaviour of cartridge fuselinks

All fuses incorporate one or more elements which melt and then vaporise when excessive currents flow through them for sufficient time, and thereafter the resulting arc or arcs must be extinguished to achieve satisfactory interruption. The means of arc extinction vary with different types of fuses. This chapter will deal with the pre-arcing behaviour of low-voltage high-breaking-capacity fuses and then variations from this process will be dealt with in later chapters when other types of fuses are being described.

2.1 General behaviour

Cartridge fuses are in many senses the most complex, containing, as they do, filling material and non-uniform elements.

Each fuselink must possess some electrical resistance, the actual amount being dependent on the configuration and material of the element or elements within it, the element/end-cap connections, the end caps and the terminals. A fuselink must therefore absorb electrical power whenever it carries current.

If the current passing through a fuselink is changed from an initial value below a certain level, either suddenly or slowly, to a steady value also below the same particular level, the temperatures of the various parts of the fuselink change, until a distribution is achieved where the heat energy dissipated from the fuselink to its surroundings and connecting cables, during each half-cycle for alternating currents or any period when direct current is flowing, is equal to the electrical-energy input in the same period. Equilibrium is thus established, and the condition could be maintained indefinitely.

If the current is increased and maintained at a value above a certain level, equilibrium cannot be achieved because, although the temperatures of the various parts of the fuselink will rise, the heat energy which will be dissipated will not become equal to the electrical-energy input by the time the whole, or parts, of the one or more fuse elements reach the melting-point temperature. As a result, disruption of the element or elements will commence and circuit interruption will follow after a period of arcing. The time from the instant when the current exceeds the critical level until the melting and vaporisation of the elements has taken place is known as the pre-arcing period and then the following time until arc interruption is achieved is called the arcing period.

The greater the amount by which the current through a fuselink exceeds the maximum level at which equilibrium can be established, the shorter is the time taken for the element or elements to reach the melting temperature. This is because the power available to cause temperature rise is equal to the difference between the power input, which is proportional to the square of the current, and the power dissipated by the fuselink. The latter quantity is dependent on the temperatures of the parts, which must be limited to the melting temperature of the element material. The power difference clearly increases with current.

As a result, all fuses have inverse pre-arcing time/current characteristics in the range above a definite minimum current level, below which they achieve equilibrium and do not operate. A typical characteristic is shown in Figure 2.1.

It is clear from this curve that there is a current for each particular fuse at which its operating time would theoretically be infinite. This current, which is the maximum at which equilibrium conditions can be attained, could not be determined experimentally because of the infinite testing time which would be involved, and for this reason the

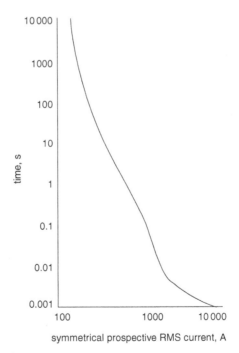

Figure 2.1   Typical time/current characteristic of a low-voltage cartridge fuselink

current needed to operate a fuse in an arbitrary, but definite, time (typically 1–4 h) is referred to in practice, as the minimum fusing current. This matter is discussed in more detail in Chapter 8.

In the following sections of this chapter the behaviour that takes place when currents of various levels are flowing will be considered separately.

2.1.1 Clearance of very high currents

When a fuselink carries a current that is very high relative to the minimum fusing current, the period required for parts or the whole of its element or elements to reach melting temperature and then go on to the end of the vaporisation stage is extremely short. In these circumstances the heat transfer both within and from the fuselink before arcing commences may be assumed to be negligible. The pre-arcing times below which this condition is satisfied vary with the type and rating of fuse, but they can be as short as fractions of a millisecond, and the corresponding currents are hundreds of times the minimum fusing current. Under these conditions of no heat transfer, the rate of temperature rise at any instant of any small piece of any component part of the fuselink carrying current may be taken to be

$$\frac{d\theta}{dt} = \frac{\text{power supplied to the piece at the instant}}{\text{mass of piece} \times \text{its specific heat}} \tag{2.1}$$

In practice, power is only supplied to the element material and thus only this material experiences the temperature rise given by the above equation. The other parts remain at their initial temperatures. The power supplied to each piece of element material is given by the square of the instantaneous current flowing through it, multiplied by the resistance it presents to the current. The latter parameter is dependent on the temperature of the particular piece and can be expressed approximately by the expression:

$$R = R_{am}(1 + \alpha\theta) \tag{2.2}$$

in which R_{am} is the resistance at the ambient temperature, α is the resistance temperature coefficient based on ambient temperature and θ is the temperature rise of the piece under consideration above the ambient value.

From eqns. 2.1 and 2.2, the following expression for the temperature above ambient at any instant may be derived:

$$\theta_t = \frac{1}{\text{mass of piece} \times \text{specific heat}} \int_0^t i^2 R_{am}(1 + \alpha\theta)\,dt \tag{2.3}$$

This equation could be solved analytically or by step-by-step numerical methods, for each piece, provided that the variation of the current through it with time was known or could be found.

There has long been a desire to calculate fuse performance so that acceptable and optimum designs can be produced without recourse to lengthy testing procedures and trial-and-error methods of modifying elements and other parts until the desired performance is obtained. In 1941, Gibson [8] published the results of investigations, done by Prof R. Rüdenberg, in which the above method was used to determine the high-current pre-arcing behaviour of fuselinks with wire elements. Rüdenberg justifiably assumed that the current flow through any element was axial and that the current density was constant, i.e. given by the total current carried by the element divided by its cross-sectional area.

For a wire element carrying an extremely high current, the behaviour of all its parts is the same, and so eqn. 2.3 may be solved for any piece to determine the time needed for the wire to reach its melting-point temperature. The times taken for the latent heats of fusion and vaporisation to be supplied, and also to provide the energy needed to raise the element temperature from its melting point to the vaporisation level, are more difficult to calculate, because the now-liquid element material may flow away from its original position. The current path is no longer so well defined and assumptions must be made, which may lead to errors. Nevertheless, an understanding of the process can be obtained and reasonably accurate pre-arcing times may be calculated. An alleviating factor is that the power losses increase considerably after melting, because of the higher resistance of the conducting path, and as a result the time taken to reach the melting point is much longer than the subsequent period ending with vaporisation, and therefore, errors in determining this latter time do not affect the overall time very significantly.

The situation is slightly more complex in fuselinks with notched-strip elements because the current density varies along the length of the element and, in consequence, the current flow is not axial at all points. Clearly the highest current densities occur in the restricted sections and these reach the melting temperature first.

It is often assumed that uniform axial current flow does take place in the restrictions, in which case the current density is given, as for a wire element, by the total element current divided by the cross-sectional area of the restriction, and eqn. 2.3 can again be used to determine the time taken for the restriction temperatures to reach the melting value. This assumption is certainly valid for elements in which the restrictions are long compared with their width. It might not, however, give accurate results for elements with short restrictions where the current flow in the restricted sections may be non-uniform because of its need to converge and diverge at the ends. In these cases one of the field-determination techniques now available could be used to determine the current distribution accurately.

In one particular study by Leach and the authors [9], the element was divided into a number of subvolumes and then, by using a finite-difference method, the current-flow patterns, one of which is shown in Figure 2.2, were determined for different instants during the period preceding melting. The heating of the element was then calculated on a step-by-step basis using eqn. 2.3 until melting and then vaporisation of a complete set of subvolumes across the element occurred, to form a break in the current path. Details of the current-pattern determination used by Leach *et al.* are given in Reference 9.

It will be appreciated that the power input varies as the square of the instantaneous current and therefore at very high currents, where the heat transfers may be neglected, the pre-arcing time of any fuselink would, with steady (direct) current, be inversely proportional to the square of the current magnitude. Such a simple relationship does

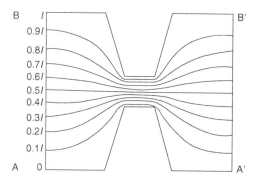

*Figure 2.2   Current-flow pattern in an element with a restricted section; I = total
           current*

not hold, however, when the current is varying with time and calculations must be
done taking account of the current variation.

After a part or parts of an element have melted, the current continues to flow
in the liquid. The situation which then arises is not fully understood but there are a
number of conditions which could be assumed to apply during the initial vaporisation
process. For the purposes of modelling the arc behaviour, Wright and Beaumont [10]
suggested the following two possible processes.

In the first, using the assumption that there is virtually no heat transfer within the
element during the short pre-arcing period associated with very high-current operation, they postulated that boiling might occur throughout each of the notched sections.
During this period there would be bubbles within the liquified element material causing the electrical resistance of the path to rise. The associated power input would then
cause rapid vaporisation of the whole of each notch. As an alternative, it could be
taken that there are temperature gradients within an element and that it will be hottest
at the centres of the restricted sections, causing vaporisation to commence at these
points. The vapour escaping through the surrounding liquid could produce a hairline
crack or gap across the notch.

The vapour in the gaps is not ionised initially and capacitance is present across
them. A similar situation must arise also with wire elements. The equivalent circuit
which would represent the conditions in these circumstances is shown in Figure 2.3.
The generator provides the source EMF (e_s) which drives current through the resistance and inductance of the source and the path up to and through the fault, including
the fuselink (R_c, R_f and L_c). The capacitance is included to allow for the breaks
in the fuselink element. Before the formation of the gaps, the capacitance would be
infinite but it would fall quickly to a low value and thereafter reduce as the gaps
lengthened, due to the further vaporisation caused by the heating effect of the current
which continues to flow. In practice, because of the relatively small cross-sectional
area of the fuselink elements, the individual capacitances would not be likely, after
a short time, to exceed a few picofarads. During the period under consideration the
resistance of the fuse element also varies.

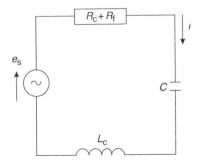

Figure 2.3   Equivalent circuit for vaporisation period

The general non-linear equation of the circuit shown in Figure 2.3 is:

$$e_s = i(R_c + R_f) + \frac{d}{dt}(L_c i) + \int_0^t \frac{i}{C} dt \qquad (2.4)$$

The current at the time ($t = 0$) when gaps are formed in the fuse element will be high, and as it cannot then fall to zero instantly, because of the circuit inductance, it will cause the voltages across the gaps to rise rapidly because of their relatively low capacitances. Breakdowns would therefore occur, after a very short period, when the voltages across each of the gaps reach a level of 17 V or less.

Because the total of the voltages across the gaps during this period will usually be small relative to the source EMF, the rate of change of the current is hardly affected by the presence of the gaps and indeed the resultant current change tends to be negligible because of the shortness of the period.

Arcs are initiated across the individual gaps when the voltage gradients across them are great enough to enable electrons to escape into the gaps from those ends of the element, which are negative in potential, i.e. those which are acting as cathodes. This occurrence may be deemed to terminate the pre-arcing period.

2.1.2   Clearance of high currents

When the currents are somewhat lower than those considered in Section 2.1.1, and the pre-arcing times are consequently longer, then it is necessary to recognise that heat movements will take place. For a certain limited range of currents, however, for which the pre-arcing times are still short, it is possible when dealing with fuselinks containing non-uniform elements, to consider that heat transfers only occur in the elements and that there are no heat movements in the other fuselink parts because of their relatively low thermal conductivities and the shortness of the periods involved.

For the pre-arcing times for which this assumption is valid, both heat conduction and storage are important and these processes are affected by the geometry of the elements. It is clear that heat must be conducted away from the restrictions, to regions where the current densities and thus the power inputs are lower, and this increases the time needed to melt and vaporise the restrictions and indeed a measure of possible control over the performance may thereby be introduced. That this is so may be seen by examining the effect of widening the sections of an element adjacent to the restrictions. Such a change lowers the current densities and power inputs in the widened sections and this in turn lowers their temperatures, causing them to draw more heat from the restrictions and thus increasing the time taken for the restrictions to melt and vaporise.

While this process can readily be understood qualitatively, the quantitative effects of changing the geometry of an element can only be determined analytically by solving equations based on energy considerations. In recent years, detailed studies have been undertaken by several workers [9, 11–14, 21]. Leach *et al.* [9] examined the behaviour of fuselinks containing notched-strip elements by first determining the current-flow pattern, as described in the previous section, and then dividing the element into a number of subvolumes as shown in Figure 2.4. Because the strip is of constant thickness and heat losses from it are being ignored, a two-dimensional study is adequate.

The heat flow by conduction between two adjacent subvolumes is clearly proportional to the temperature difference between their centres and the cross-sectional area for heat flow, and it is inversely proportional to the distance between their centres.

The heat flow by conduction in a time Δt into the subvolume mn, shown in Figure 2.4, is therefore given by:

$$\Delta h_C = ZK\Delta t \left\{ \frac{(\theta_{m-1,n,} - \theta_{m,n})Y_n}{\frac{1}{2}(X_{m-1} + X_m)} + \frac{(\theta_{m,n-1} - \theta_{m,n})X_m}{\frac{1}{2}(Y_{n-1} + Y_n)} + \frac{(\theta_{m+1,n,} - \theta_{m,n})Y_n}{\frac{1}{2}(X_m + X_{m+1})} \right.$$
$$\left. + \frac{(\theta_{m,n+1} - \theta_{m,n})X_m}{\frac{1}{2}(Y_n + Y_{n+1})} \right\} \tag{2.5}$$

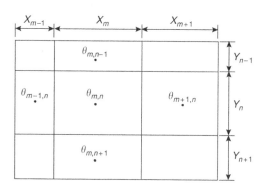

Figure 2.4   *Five adjacent subvolumes*

Where
D = density
K = thermal conductivity
Z = thickness
λ = specific heat

in which $\theta_{m,n}$, $\theta_{m-1,n}$, $\theta_{m,n-1}$, $\theta_{m+1,n}$ and $\theta_{m,n+1}$ are the mean temperatures of the subvolumes during the interval, i.e. the temperatures at the centres halfway through the interval.

The increase in heat energy stored by the subvolume mn in time Δt, if its average temperature at the end of the interval prior to that under consideration was $\theta'_{m,n}$, is given by

$$\Delta h_s = 2(\theta_{m,n} - \theta'_{m,n})\lambda D X_m Y_n Z \tag{2.6}$$

It is evident that the sum of the heat flow into any subvolume and the heat generated within it must be equal to the increase in heat energy stored by it, i.e.

$$\Delta h_g + \Delta h_C = \Delta h_S \tag{2.7}$$

where Δh_g is the heat generated in the subvolume.

Substitution from eqns. 2.5 and 2.6 in eqn. 2.7 enables $\theta_{m,n}$ to be expressed in terms of the mean temperatures of the surrounding subvolume as follows:

$$\theta_{m,n} = \frac{\frac{1}{2}\Delta h_g + ZK\Delta t(P\theta_{m-1,n} + Q\theta_{m,n-1} + R\theta_{m+1,n} + S\theta_{m,n+1}) + \theta_{m,n}\lambda D X_m Y_n Z}{\lambda D X_m Y_n Z + ZK\Delta t(P + Q + R + S)} \tag{2.8}$$

in which

$$P = \frac{Y_n}{X_{m-1} + X_m}, \quad Q = \frac{X_m}{Y_{n-1} + Y_n}, \quad R = \frac{Y_n}{X_m + X_{m+1}}, \quad S = \frac{X_m}{Y_n + Y_{n+1}}$$

A method of successive over-relaxation similar to that suggested earlier to determine the current distribution may be used to solve eqn. 2.8 and thus give the temperature distributions at the middle of each time interval.

The final temperatures of the subvolumes at the ends of the time intervals are obtainable from equations of the form:

$$\theta''_{m,n} = 2\theta_{m,n} - \theta'_{m,n} \tag{2.9}$$

in which $\theta''_{m,n}$ is the temperature of the centre of subvolume mn at the end of the interval under consideration.

From the final temperatures of the subvolumes, the resistivities needed for the next step in the calculation may be obtained. The modified current distribution which results from the resistance changes may then be calculated and the energy inputs for the next step can be determined. In this way the computations can be continued until melting and then vaporisation of the subvolumes in the restrictions occurs.

2.1.3 Behaviour at intermediate current levels

For currents somewhat lower than those considered in the previous section and for which the pre-arcing times are longer, heat movement is not confined to the elements. Certainly, in any analysis, allowance must be made for heat movement out of the element to the surrounding filling material and body, and possibly also to the terminals and connecting cables.

It will be appreciated that account should be taken of the three-dimensional nature of both the temperature distribution and the heat movements, if accurate computational studies are to be undertaken and, of course, the geometry is quite complex making it necessary to divide up the fuselink into a very large number of subvolumes of varying sizes. This inevitably increases the computation needed for each time interval in step-by-step studies.

To simplify the calculations for medium pre-arcing times Leach *et al.* [9] assumed that the heat transferred from an element to the filler flowed only in directions normal to the main surfaces of the element, and that there is no longitudinal heat flow in the filling material. These assumptions are not particularly valid for short fuses of large diameter, but they do appear to be very reasonable for fuselinks which have a relatively high ratio of length to diameter. Even with these assumptions, both the element and the filler must be divided into subvolumes to allow heat movement in the x, y and z directions.

2.1.4   Behaviour at currents near the minimum fusing level

The treatment described above could be further extended to deal with longer pre-arcing times but the added complexity needed to allow for all the heat movements would be great and, because the computing time would be unacceptably long, an alternative approach is necessary.

As stated earlier, there is a current for every fuselink at which melting would theoretically commence after an infinite time, and time/current characteristics therefore become asymptotic to a line drawn at this current, which is approximately the minimum-fusing-current level. Under these conditions, the temperature and resistivity changes are so slow that the final situation may be assumed to be a steady state in which no changes in stored energy occur. Leach *et al.* [9] analysed this equilibrium condition and determined the current at which the restrictions in the elements of the fuselinks under consideration were at the melting-point temperature.

When doing these calculations, allowance must be made for heat movements not only within the fuselink being considered but also in its surroundings and the connections to it. This may be done, as before, by dividing the various parts into subvolumes and calculating on a full three-dimensional basis. For equilibrium to be maintained, the heat energy generated within each subvolume and the heat flow to it must be equal to the energy dissipated from it in any period, that is there must be no change in stored energy. The analysis must allow satisfactorily for heat transfer at boundaries between materials with different thermal conductivities. This can be done by considering two adjacent subvolumes as shown in Figure 2.5a and assuming uniform temperature gradients between the centre of each subvolume and its boundaries as shown in Figure 2.5b. It can be seen that the heat flow between the subvolumes is given by

heat flow from subvolume (m, n) to $(m + 1, n)$ in time Δt

$$= 2\frac{\theta_{m,n} - \theta}{X_{m,n}} Y_n Z K_{m,n} \Delta t$$

in which θ is the temperature of the boundary.

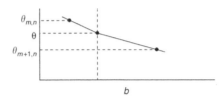

Figure 2.5   Heat flow between subvolumes of dissimilar material

Where
K_{mn} = thermal conductivity of subvolume m, n
$K_{m+1,n}$ = thermal conductivity of subvolume $m + 1, n$
Z = thickness

heat flow into subvolume $(m + 1, n)$ from (m, n) in time Δt

$$= 2\frac{\theta - \theta_{m+1,n}}{X_{m+1,n}}Y_n Z K_{m+1,n}\Delta t$$

These heat flows must be the same and by equating them, θ may be eliminated and the heat flow may be expressed in the form

$$\text{heat flow} = \frac{2K_{m,n}K_{m+1,n}(\theta_{m,n} - \theta_{m+1,n})Y_n Z\Delta t}{K_{m+1,n}X_m + K_{m,n}X_{m+1}} \qquad (2.10)$$

When heat flows through a short path in a subvolume of high thermal conductivity into a longer path in a subvolume of low conductivity, the above expression (eqn. 2.10) may be simplified to

$$\text{heat flow} = \frac{2K_{m,n}(\theta_{m,n} - \theta_{m+1,n})Y_n Z\Delta t}{X_m} \qquad (2.11)$$

for $K_{m+1,n} \gg K_{m,n}$ and $X_m > X_{m+1}$.

This situation applies, for example, at the boundaries between a fuse body and its end caps and between the elements and filler, and this simpler expression may be used at appropriate boundary points to reduce the amount of computation.

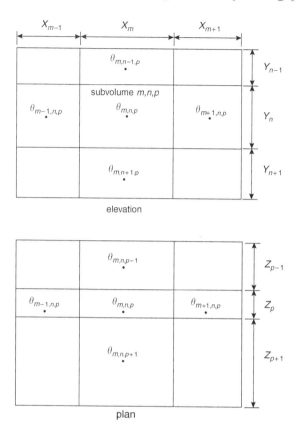

Figure 2.6   Seven adjacent subvolumes

Where

$K_{m,n,p}$, $K_{m,n,p+1}$ = thermal conductivities of subvolumes m, n, p; m, n, $p+1$ etc.

In examining the heat energy conducted into any subvolume three-dimensionally, the six subvolumes adjacent to it, and shown in Figure 2.6, must be considered: total heat conducted into subvolume m, n, p

$$
= \Delta h_{\mathrm{c}} = \frac{2K_{m+1,n,p}K_{m,n,p}(\theta_{m+1,n,p} - \theta_{m,n,p})Y_n Z_p \Delta t}{K_{m+1,n,p}X_m + K_{m,n,p}X_{m+1}}
$$
$$
+ \frac{2K_{m,n-1,p}K_{m,n,p}(\theta_{m,n-1,p} - \theta_{m,n,p})X_m Z_p \Delta t}{K_{m,n-1,p}Y_n + K_{m,n,p}Y_{n-1}} + \cdots
$$
$$
+ \frac{2K_{m,n,p-1}K_{m,n,p}(\theta_{m,n,p-1} - \theta_{m,n,p})X_m Y_n \Delta t}{K_{m,n,p-1}Z_p + K_{m,n,p}Z_{p-1}} \tag{2.12}
$$

During any time interval, the sum of this energy and that generated within the subvolume (Δh_{g}) must be equal to the heat lost from it by convection and radiation (Δh_1), i.e.

$$\Delta h_g + \Delta h_c = \Delta h_1 \tag{2.13}$$

From eqns. 2.12 and 2.13, the following expression can be derived for the temperatures of the subvolumes:

$$\theta_{m,n,p} = \frac{\dfrac{\theta_{m+1,n,p} K_{m+1,n,p} Y_n Z_p}{K_{m+1,n,p} X_m + K_{m,n,p} X_{m+1}} + \dfrac{\theta_{m,n,p-1} K_{m,n,p-1} X_m Y_n}{K_{m,n,p-1} Z_p + K_{m,n,p} Z_{p-1}} + \cdots + \dfrac{\Delta h_g}{\Delta t} - \dfrac{\Delta h_1}{\Delta t}}{2 K_{m,n,p} \left(\dfrac{K_{m+1,n,p} Y_n Z_p}{K_{m+1,n,p} X_m + K_{m,n,p} X_{m+1}} + \cdots + \dfrac{K_{m,n,p-1} X_m Y_n}{K_{m,n,p-1} Z_p + K_{m,n,p} Z_{p-1}} \right)} \tag{2.14}$$

It will be appreciated that the Δh_g term is zero except for subvolumes which carry current, and the Δh_1 term is zero except for subvolumes which lie on the fuselink surfaces. When considering the latter, one or more of the thermal conductivity terms is zero.

For a particular current, an iterative process may be used to determine the steady-state temperature distribution, adjustments being made to allow for the resistivity and current distributions. This calculation may be repeated to determine the current level at which the fuse-element restrictions are at their melting-point temperature. The same process may also be used to determine the steady-state running conditions at currents below the minimum fusing level.

2.1.5 Mathematical and experimental studies

Leach *et al.* [9] used the methods described in the preceding sections to calculate the performances of existing fuselinks so that the accuracy of the results given by them could be assessed. In addition, the ranges over which the various heat-flow assumptions are reasonably valid were studied.

The determination of current-flow patterns within fuse elements forms a fundamental part of the analytical technique and therefore a study was made of the distributions in a number of elements, each assumed to be operating with a uniform electrical resistivity, the value chosen being that of silver at 20°C. It is evident that the resistance of a fuse element can be readily calculated once the current-flow pattern is established. This was done for each of the elements considered, and close agreement was obtained, thus verifying the calculations.

In subsequent calculations, the current-flow patterns were determined at various stages during pre-arcing periods and it was found that the patterns do not change greatly as the elements heat up although of course the overall resistance of an element does rise significantly and the energy inputs rise very greatly in the restricted sections where the resistivities increase to high levels as melting temperatures are approached. Figure 2.7 illustrates the smallness of the change in the current distribution in a typical element between the beginning and the end of the pre-arcing period. The initial distribution (i.e. when cold) is shown in solid lines and the distribution as the restrictions near their melting point temperature is in broken lines.

Pre-arcing times were calculated at extremely high currents for various element forms, in which it is assumed that there is no heat movement, using the method

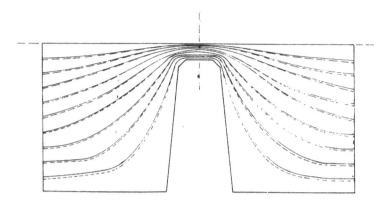

*Figure 2.7   Current-flow lines for uniform and non-uniform temperature distribu-
tions*

—————— for uniform temperature distribution
— — — — — for non-uniform temperature distribution

described in Section 2.1.1. Comparison of the results obtained with those determined
experimentally showed that there is close agreement for fuselinks with plain cylindrical wire elements at currents which give pre-arcing times up to 35 ms. With many
modern industrial fuselinks (e.g. those which comply with *IEC 60269-2-1*), close
agreement was only obtained for currents at and above those which give pre-arcing
times of about 3 ms, and the situation was even more restricted for fuselinks used
to protect semiconductors, satisfactory agreement only being obtained for currents
which gave pre-arcing times of 0·3 ms or less. This led to the interesting conclusion
that, although in the past the simple treatment, which is possible when heat movement
is neglected, was satisfactory for a significant part of the time/current characteristic,
this is no longer so, and for many present-day fuselinks it is only applicable for severe
short-circuit currents.

Pre-arcing times were calculated allowing for heat movement in the fuselink
element but not in the other parts. Close agreement was obtained with experimentally
determined values for industrial and semiconductor fuselinks for conditions leading
to pre-arcing times up to 5 and 10 ms, respectively.

Pre-arcing times were also determined by calculation assuming heat movement
to occur in the fuse elements and the filling material. A calculated characteristic
for a semiconductor-protection fuselink, together with the associated experimentally
obtained values, is shown in Figure 2.8. It can be seen that the curve begins to diverge
for this fuselink at currents which give pre-arcing times of 5–10 s, indicating that the
effects of heat losses from the fuselink begin to become significant at times of this
order.

An interesting effect which was revealed during the calculations was a disconti-
nuity in the time/current characteristic for sinusoidally varying currents, that is for
currents of the form $i = I_{pk} \sin \omega t$. This effect had been noted in the past in experimental results but its cause had not been understood. During periods when the magnitude

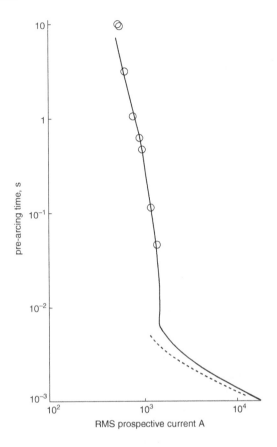

Figure 2.8   Pre-arcing time/current curve for a 200A semiconductor-protection fuselink carrying 50 Hz symmetrical sinusoidal current

      ——————   predicted results
      o o o   test results
      — — — —   curve from adiabatic formula
      —·—·—   discontinuity discussed in Section 2.1.5

of the current is rising the element heats up rapidly and, if vaporisation does not occur before the current reaches its maximum value, the element temperatures fall during the following quarter cycle as the energy input reduces and heat is conducted away. This behaviour is particularly pronounced with fuselinks which incorporate elements with restricted sections. The maximum temperatures are attained somewhat after the instants of peak current because of the heat conduction. For the fuselinks which were studied in detail, the maximum temperatures were reached between 6 and 7 ms after current zeros (for 50 Hz sinusoidal currents). Fuse operation is only possible in these periods and if it does not occur in them it will not take place until the current is rising again. The effect occurs with over-currents of levels which cause operation in one or two half-cycles, examples being shown in Figure 2.9. Curves of

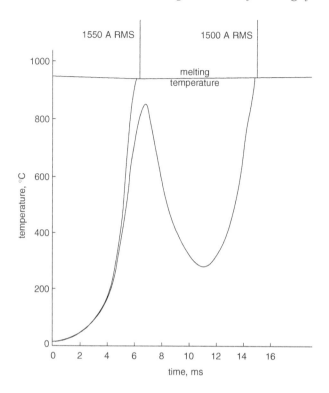

Figure 2.9   Restriction-temperature/time curves for a semiconductor-protection fuselink carrying two symmetrical sinusoidal currents

the fuse-element-restriction temperature variation with time are shown for two symmetrical fault currents, one just causing operation within the first half-cycle (6·5 ms) and the other, with only a 3 per cent lower RMS value, causing operation in the second half-cycle (15 ms). Confirmation was obtained from practical tests and, with the fuselink for which the calculations had been made, operation was not obtained between 7 and 14 ms after the application of current. It will be clear that operation in the time range up to about 40 ms is dependent on the current waveform and, for waves which are basically sinusoidal, the behaviour is affected by the instant in the supply-voltage cycle at which a fault occurs and also by the circuit parameters as these affect the transient component of the current wave. At lower currents, for which the operating times are much longer, it is only the RMS value of the current which effectively controls the behaviour and the instantaneous variations are not normally taken into account in theoretical studies.

The temperature distributions in fuselinks can readily be obtained mathematically using the equations given in earlier sections and such information, which cannot be determined by other means, is valuable to designers and application engineers. Calculated distributions over a section of the element of a semiconductor fuselink rated at 200 A when carrying sinusoidal currents of 900 and 1200 A are shown in

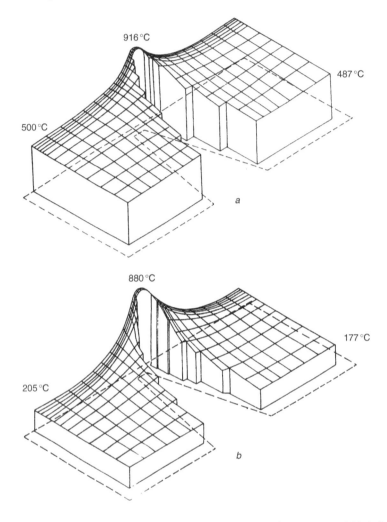

Figure 2.10   Temperature distribution, just prior to melting over a 200 A fuselink element at two current levels
   a   900 A RMS
   b   1200 A RMS

Figure 2.10 in which temperature is proportional to height. Both the distributions shown apply at times just prior to the melting of the restrictions, the pre-arcing times for the two conditions being 0·7 and 0·11 s. It is interesting to note the generally lower temperatures of the majority of the element at the higher current, this being due to the lower conduction of heat from the restriction because of the shortness of the pre-arcing period.

In addition to the above studies, calculations were done in the manner suggested in Section 2.1.4 to determine the minimum currents at which various fuselinks would operate, i.e. after an infinite time, and these corresponded well with values obtained experimentally.

Calculations of temperature distributions at rated currents were also done and the effects on them of the cross-sectional and surface areas of the connecting cables were studied and reported on by Leach *et al.* [9].

Finite-difference methods such as the one described here are often formulated in terms of an equivalent thermal resistance–capacitance network (Beaujean *et al.* [15]), which allows standard circuit analysis software to be used to obtain the temperature distributions. However, practical fuse designs are truly three-dimensional, with multiple parallel notched elements. A very large and complex mesh of subvolumes is required, with very small subvolumes in the notch zones, and larger subvolumes in the outer regions of the fuse body and terminals. Simplifying assumptions are needed to obtain solutions economically.

Finite-element methods have also been used to model fuse thermal behaviour, but again the models have usually been very much simplified to allow solution with a standard finite-element, FEA, package. However, Kawase *et al.* [16] wrote a finite-element program to give a true three-dimensional transient model of a semiconductor fuse with multiple parallel notched elements. Although excellent results were obtained, the computing time was excessive. A significant increase in computing power is required before FEA analysis can be used as a routine design tool.

2.2   Control of time/current characteristics

The previous sections have been included because it is felt that they enable a clear understanding to be obtained of the factors which influence the various sections of time/current characteristics.

It is clear that the minimum operating current level to which the time/current characteristic becomes asymptotic depends not only on the complete fuselink and its housing but also on the connecting cables because equilibrium must be established at all points for the condition to be maintained indefinitely. Clearly the use of larger cables or an increase in the surface area of a fuselink enables the heat dissipation to be increased with the resultant rise in minimum operating current.

At somewhat higher current levels, the performance is no longer significantly affected by the heat dissipated by the body and the connections but only by the material and geometry of the element and the properties and packing density of the filler.

At still higher current levels, the behaviour is only dependent on the element and at extremely high currents, where heat movement is negligible, the pre-arcing time is entirely controlled by the element material and the dimensions of the restrictions.

Because certain physical parameters principally affect different sections of the characteristics it should, within limits, be possible to obtain the required short-circuit performance by suitable restriction design and then the other parts can be controlled in turn by choosing the appropriate element geometry, filling, body and end-cap design and finally the housing and connections.

2.3   M-effect

In 1939 A. W. Metcalf published articles under the title 'A new fuse phenomena' [18]. They concerned the effect of the amalgamation of metals on fuse performance. It had been found, after carrying out work on silver fuse elements with ordinary tinman's solder, that the solder appeared to act on the silver at temperatures above the normal operating level, and this sometimes caused the element to melt at the position of the solder rather than at more restricted sections where the maximum temperatures would be expected.

To conduct an initial study of the phenomenon, a large number of reference fuses were made, each consisting simply of a length of silver wire housed in a glass tube which was filled with sand and sealed at the ends. All the fuses were the same except that, whereas one group contained only a plain simple wire, the others had a globule of solder on the centre of the wire, as shown in Figure 2.11. The current/operating time curves obtained by Metcalf for these fuses are shown in Figure 2.12, from which it can be seen that the minimum fusing current for the fuses with the solder globules was only about 60 per cent of that of the plain fuses. It was deduced that the temperature rise on melting of the fuses with the globules of silver was only about 345°C, that is $(0.6)^2$ times the melting point of silver, it being assumed that the heating effect was proportional to the square of the current.

During further tests in which the current was raised in small increments, steady temperatures being allowed to develop after each increment, a change was observed when the globules melted. This was followed by progressive changes in the appearance of the globules which soon became red hot in spite of the fact that the wires adjacent to them remained unchanged in appearance. The globules only glowed red for a very short time and clearance was effected by the silver wires rupturing immediately adjacent to the globules. Subsequent examinations showed that the silver wire

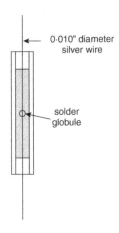

0·010" diameter
silver wire

solder
globule

Figure 2.11   Reference fuse produced by A. W. Metcalf

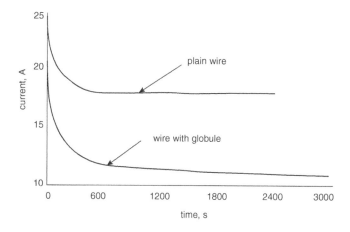

Figure 2.12   Current/operating time curves showing the effect of solder on a silver-wire fuse element

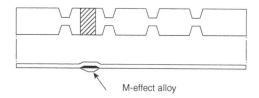

Figure 2.13   Fuselink element with M-effect alloy

in contact with solder had entirely diffused into it, and the wire where it entered a globule was very brittle and broke away when touched.

These experiments were repeated, but on this occasion the currents were interrupted before the wires melted. It was found that the resistances of the wires passing through the globules had increased by 100 per cent whereas the wires in fuses without globules showed no measurable increase in resistance on cooling.

This phenomenon is now exploited extensively and many fuselinks have low-melting-point metals deposited on the main element material which is usually of silver or copper. When the element is in the form of a strip with restricted sections the low-melting-point metal is deposited adjacent to the restrictions but not always on them, as shown in Figure 2.13. For overload protection, at currents of sufficient magnitude to cause the low-melting-point material to melt, the alloying process commences, the element resistance rises in the alloy region and rupture occurs as described above. The current for operation is lower than that which would be required in the absence of the low-melting-point metal and lower fusing factors can be obtained. At the low operating currents the restrictions do not melt.

At the very high currents associated with short circuits, the restrictions heat up very rapidly and reach their melting point temperatures after only a few milliseconds. The temperatures at the points where the low-melting-point material is deposited do not reach the level needed for the alloying process to commence because of the thermal mass of these sections and the insignificant heat transfer from the restrictions during the short operating times.

It is thus possible, by choosing suitable materials and dimensions, to obtain time/current characteristics which could not be produced in fuselinks containing elements of only one material.

It will be appreciated from the above that the techniques described earlier in Section 2.1 may be used to calculate the operating times of fuselinks with elements incorporating M-effect materials when they carry currents of many times their minimum fusing levels. At lower current levels, however, modelling techniques must take into account the diffusion of the low-melting-point material into the remainder of the element.

Recently Beaujean *et al.* [19] produced a model capable of determining, on a step-by-step basis, the current and temperature distributions in an element and also the diffusion of the low-melting-point material into the main element material. The latter process was modelled using Fick's second law of diffusion, i.e.

$$\frac{\partial c}{\partial t} = D \left(\frac{\partial^2 c}{\partial x^2} + \frac{\partial^2 c}{\partial y^2} + \frac{\partial^2 c}{\partial z^2} \right)$$

in which $\partial c / \partial t$ is the rate of change of the solute concentration at a point with time, x, y and z are the displacements of the point in the three dimensions and D is the diffusion coefficient for the pair of materials.

The diffusion coefficient D does not have a constant value for a given pair of materials, but varies with both temperature and solute concentration, and it is clearly necessary that these relationships are known when modelling of behaviour is undertaken.

A further factor, which must be taken into account, is that the current densities are not constant across the thickness of an element at positions where the low-melting-point material is present and the variations in current density at such positions change as the diffusion progresses. As a consequence, the current distributions cannot be determined at each time step on the two-dimensional basis employed by Leach *et al.* [9] but must be done three dimensionally.

Beaujean and Newbery therefore determined the current and temperature distributions and then the solute concentrations in an element, on a three-dimensional basis at each time step. They also performed experiments to determine the rates at which tin diffuses into silver for a range of conditions. Using this information, they computed the time/current characteristics of fuselinks containing elements both with and without restrictions. Figure 2.14 shows two of the computed characteristics which they obtained, together with experimentally measured characteristics. These indicate that reasonably good agreement was obtained and, with further improvements in such techniques, it should be possible to develop models which will enable fuselinks employing the M-effect to be produced with the characteristics needed for particular applications without recourse to the lengthy testing procedures associated with trial-and-error methods.

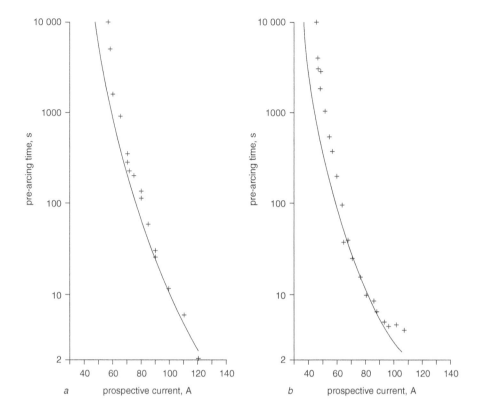

Figure 2.14 Time/current characteristics for elements without restrictions (a) and with restrictions (b)

Mathematical modelling of the M-effect is very complicated, involving the solution of three coupled field problems: electrical, thermal and diffusion. A subject review has been given by Lindmayer [17], who also devised an original numerical method to obtain solutions of the three coupled field problems for a simple fuse design.

2.4 Skin and proximity effects

Fuselinks with high current ratings often contain several identical elements connected in parallel. In this situation it is intended that the elements should share the current equally, in which event the pre-arcing performance can be determined as described earlier.

In some applications, however, the sharing may not be equal and it is important that this should be recognised and the necessary allowances made to ensure that the protection provided by the fuselinks is satisfactory. Two possible causes of unequal sharing are skin effect and the presence of current-carrying conductors near

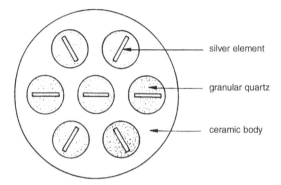

Figure 2.15   Fuselink with seven elements

Figure 2.16   Representation of strip elements by circular conductors

the fuselinks (proximity effect). These factors were examined by Howe and Jordan [20] and their findings are summarised below.

They did a theoretical study of a fuselink with seven, parallel strip elements of uniform cross-section arranged as shown in Figure 2.15. For modelling purposes, each element was represented by five parallel circular conductors as shown in Figure 2.16. The current distribution was calculated assuming the return current path to be very remote and, as would be expected, the current densities increased with radial distance from the centre of the fuselink, the variation increasing with frequency. This is clearly the behaviour to be expected because of the skin effect. Because of the symmetry of the arrangement, each of the outer elements carried the same current but the central element carried a current which was different in phase and magnitude. This is illustrated in Figure 2.17 which shows the locus of the current in the centre element, allowing for some circuit inductance, and taking the current in each outer element as the phasor reference and of one unit in magnitude. This shows that the effect is very small at power frequencies but makes it clear that the current in the centre element

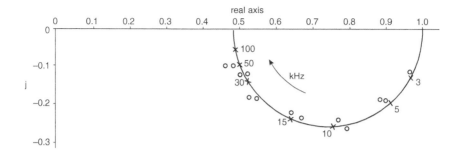

Figure 2.17 *Locus of the current in centre element with respect to the current in an outer element*

× theoretical points
o practical points
Frequencies are in kilohertz

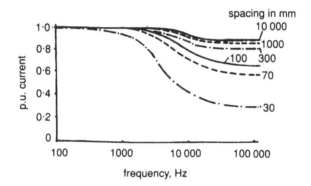

Figure 2.18 *Fuselink rating/frequency characteristics for various spacings from the return conductor*

can fall to a low level at very high frequencies. Howe and Jordan showed that, in the limit, the ratio can be as low as 0·37 and this would result in the outer elements carrying 10 per cent more than they would at very low frequencies.

The proximity of a return or other current-carrying conductors near to multi-element fuselinks may be significant at high frequencies. In general, the effect of a return conductor is to cause the currents in the elements nearest to it to increase and consequently, the more remote elements carry lower currents. The result is again, a non-uniform current distribution in the elements.

It is very important that the current rating of an individual element of a fuselink is not exceeded for long periods because this could cause it to rupture. Thereafter the current it was formerly carrying would be diverted through other elements causing them to melt in turn to give eventual clearance of the fuselink. To prevent this occurring due to skin effect and/or proximity effect, it is necessary to reduce the current ratings of fuselinks which are to be used at high frequencies. Figure 2.18 was derived by

Howe and Jordan as the derating characteristic for a fuselink with elements arranged as shown in Figure 2.15.

It will be appreciated that the skin effect, but not the proximity effect, can be overcome by mounting all the elements at the same radius from the fuselink centre and this is therefore a desirable construction for fuses which are to carry high-frequency currents.

High-frequency applications are becoming increasingly common because of the introduction of power electronic equipment, for example, invertors using fast switching power transistors in AC drives, UPS systems and induction heating applications (see Section 7.8.8). Such equipment operates at frequencies above 10 kHz and experience has shown that fuselinks used with such equipment do maloperate unless allowance is made for the above effects.

Chapter 3

Arcing behaviour of cartridge fuselinks

The previous chapter dealt with fuselink behaviour during the pre-arcing period which ends when breaks are formed in the element because of parts of it running away in liquid form or vaporising.

It was stated in Section 2.1.1 that there must be a very short period during which voltages build up across the breaks and these lead to ionisation and the formation of arcs. The arcs then persist until the current reaches zero, at which time arc extinction occurs, and clearly it is desirable that interruption should be maintained and that the arcs should not restrike. This process must always occur and it is important that satisfactory clearance is achieved at all current levels. At currents in a limited range above the minimum fusing value, however, the duration of the period in which arcing takes place is very short relative to the total clearance time and the effects of arcing on such factors as the energy let-through to the circuit being protected are usually insignificant and therefore not of interest. In these circumstances it is possible to neglect the arcing period and predict the behaviour and operating times by only considering the pre-arcing performance as outlined in Chapter 2.

This is not possible, however, when the behaviour at very high currents is being considered because the period for which arcing persists may well be comparable with or even greater than the pre-arcing period and the energy let-through to the protected circuit during the arcing period represents a considerable fraction of the total.

To determine the arcing-period duration, it is necessary to predict the current variation from the end of the pre-arcing period and to thus determine when the current will fall to zero. To do this, the relationship between the current through the fuselink and the voltage across it during arcing must be known or be calculable as can be seen from the following section.

3.1 Basic conditions during the arcing period

During the pre-arcing period the current will have varied with time in a manner determined by the source EMF and the impedances in the circuit up to the point of

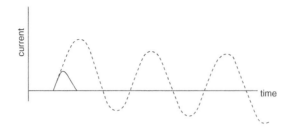

Figure 3.1   Prospective and actual currents

        –––––––– prospective current
        ––––––––– actual current

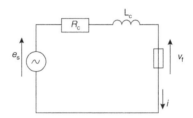

Figure 3.2   Equivalent circuit

fault. It will be slightly lower at the end of the period because of the fuse resistance than it would have been had the fuse not been present. The current which would have flowed in these circumstances is known as the prospective current and typical variations are shown in Figure 3.1. The lower the circuit impedance, the greater is the reduction of current caused by the presence of the fuse, and the degree of current limiting is thus greatest at very high prospective currents. This in itself is a very advantageous feature, as the energy let-through to the circuit protected by the fuselink is thereby considerably reduced during the most severe fault conditions.

At the instant when arcs are initiated in a fuselink, there is a significant increase in the voltage drop across it. This voltage then rises as the arcs lengthen owing to more metal being eroded from the element, because of the high arc temperature.

Consideration of the simple circuit shown in Figure 3.2, which is assumed to apply for a fault condition, shows that the basic voltage/current relationship is:

$$e_s = i R_c + \frac{d}{dt}(L_c i) + v_f \qquad (3.1)$$

in which

e_s = source EMF
R_c = resistance of circuit except for fuselink
L_c = inductance of circuit

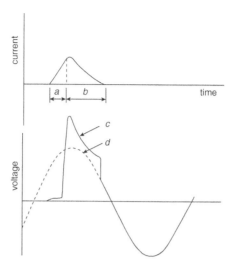

Figure 3.3   Electrical conditions during short circuit

Where
a   pre-arcing period
b   arcing period
c   fuselink voltage
d   source EMF

v_f = voltage across the fuselink including the resistance drop within it
i = circuit current.

For the condition in which the current is positive at the commencement of arcing, it is necessary for the rate of change of current (di/dt) to become negative so that the current will fall to zero, allowing the arcs to extinguish. This occurs, when ($e_s - iR_c - v_f$) is negative. Clearly although this condition may not be met at the start of arcing and the current may continue to rise, it will eventually be met and the current will then fall. To achieve rapid extinction it is necessary that the voltage across the fuse should be large, as this causes the current to begin to fall earlier and to reach zero more quickly. These conditions are illustrated in Figure 3.3.

Fuselinks containing notched-strip elements may be made to reduce the current more rapidly during the arcing period by increasing the number of restrictions, because this leads to the establishment of several arcs in series and consequently a greater voltage across the fuselink. This must, however, be limited to a level which will not cause such a large rate of change of current that excessive voltages may be induced in inductive components, and upper limits for fuselinks voltages are quoted in specifications. Although it cannot be achieved, the ideal situation would be for the fuselink voltage to rise to the limiting value at the commencement of arcing and to remain there until current interruption is achieved. This would give the fastest possible fault clearance and thus reduce to a minimum the amount of energy supplied to the protected circuit.

It should be appreciated that the situation is much less controlled in fuselinks with cylindrical-wire elements. Theoretically the whole of such an element should have uniform conditions in it, but in practice element distortion occurs during the pre-arcing period, as a result of which the cross-sectional area does not remain constant and there are variations along the length, producing bulges, i.e. thinner and fatter sections, known as unduloids. Gaps ultimately form at the centres of the thin sections and their number is not fixed, as with notched-strip elements. In some cases, the number of arcs may be great enough to cause excessive fuselink voltages to be produced.

Because of the insignificant heat movements which occur during the clearance of very high currents, the thermal properties of the body, end caps and terminals of a fuselink have no significant effect on the performance or arcing, provided that they are not affected by thermal or mechanical shock. It is the element material and configuration, particularly its restrictions, and the characteristics of the filling material which are of prime importance.

Empirical relationships of fuselink voltage and current during arcing have been produced to enable eqn. 3.1 to be solved to determine the current variations for various applications. A relationship developed by Gnanalingam and Wilkins [21] enabled them to predict the performances of certain fuses with reasonable accuracy and they claimed that the simulation technique is useful for screening preliminary designs and for investigating the effects of various system parameters such as frequency.

Such methods do not, however, deal with the underlying phenomena and are not of much assistance in explaining behaviour, developing fuses or considering the suitability of materials different to those which are presently used. It must be recognised that the arcing process is very complex and it is therefore unlikely that a completely accurate model can be constructed. Nevertheless, Wright and Beaumont [10] did develop a mathematical model, admittedly based on a number of simplifying assumptions, which they considered to be superior to earlier models in that it did attempt to deal with the detailed processes involved during arcing. It was therefore felt to be more general than other models and apart from aiding understanding it was hoped that it would enable the effects of changes in materials and parameters to be predicted and thus enable designs to be optimised. It probably must be accepted that it will always be necessary to determine fuselink characteristics by experimental means.

The development of the modelling technique of Wright and Beaumont is outlined in the following sections.

3.2   Arc model

Arcing is assumed to commence when a gap in the fuselink element becomes ionised because of the rapid buildup of voltage across it by the process described in Section 2.1.1. Thereafter, calculation may proceed on a step-by-step basis to determine the voltage and current variations.

It is common to regard an arc as having three main regions, i.e. the cathode-fall region, the anode-fall region and the positive column. These regions, which each have their own properties, are considered in the following subsections and they may be dealt with separately in the model.

3.2.1   Cathode-fall region

At one end of this region, which is only about 10^{-3} mm long, the current flow is in metallic vapour whereas at the other end it is in the solid or liquid metal cathode.

Electron emission from low-boiling-point cathodes, such as silver, can be explained by a thermal and field mechanism described by Dolan and Dyke [22], which takes account of the combined effects of the temperature of the cathode spot and the electric field at the spot. The voltage drop associated with the cathode-fall region is found to be substantially constant at about 10 V and this value was assumed by Wright and Beaumont for their model.

3.2.2   Anode-fall region

This region is also about 10^{-3} mm long and, as with the cathode-fall region, transition occurs within it whereby the current flow is in the solid or liquid metal anode at one end and in the metal vapour at the other.

Unlike the situation in the cathode-fall region, however, where the voltage drop is almost constant, the voltage across the anode-fall region (V_{af}) could be of any value between almost zero and the ionisation potential of the atoms of the element material. The actual voltage depends on the mechanism prevailing in the region [23]. A constant voltage equal to the ionisation potential (7·56 V for silver) is probably the most reasonable value to use in a model because field ionisation is the probable mechanism for ion production in fuse arcs.

3.3   Positive column

The positive column occupies the space between the cathode- and anode-fall regions and charge neutrality exists within it. In practice the conditions within a column are non-uniform, but to simplify modelling, Wright and Beaumont neglected the variations and assumed the column to have, at any instant, a fixed cross-sectional area throughout its length and also the same conductivity at all points. With these assumptions its resistance can clearly be calculated provided that its effective dimensions and conductivity can be determined. These parameters may be considered and determined as described in the following subsections.

3.3.1   Length of a positive column

As stated earlier, the electrode-fall regions in an arc are extremely short and unlikely to be more than 10^{-3} mm long and therefore the positive column may be taken to occupy the whole gap length between the electrodes. This length is dependent on the power inputs and interchanges which cause erosion of the element and therefore extension of the gap.

3.3.1.1   Power supplied to an anode

An anode receives significant power in the following two ways:

(a) The power used in a cathode to cause electrons to leave its surface is returned to an anode when the electrons reach and enter it. The associated power is equal to the product of the work function (a voltage V_{wf}) and the current flowing into the anode.

(b) The kinetic energy of the electrons reaching an anode is given up to it. The associated power is made up of two parts, i.e. the thermal power acquired in the column, equivalent to the electrons falling through a voltage drop (V_T) of 1 V [24] and that due to the acceleration experienced by the electrons in the anode-fall region.

Power is also given to an anode in other ways including the following:

(c) Radiation and conduction from the positive column. From calculations which have been done, it appears that with the areas and at the temperature differences which exist in practice the power levels associated with these modes of heat transfer are very small compared with those referred to under (a) and (b) above.

(d) Joule heating of the anode. Assessments of this effect have also been made [10] and these indicate that the associated power is likely to be small. This confirms the view expressed earlier by other workers.

3.3.1.2   Power lost by an anode

An anode dissipates power in several ways including radiation and heat conduction into the surrounding filling material and also away from the arc into adjacent parts of the element. The levels tend to be small in each case and an example is given in Reference 10 to illustrate the magnitude of the heat conducted along the element.

3.3.1.3   Power supplied to a cathode

A cathode receives significant power in the following ways:

(a) The ions reaching a cathode use some of their energy in overcoming the surface forces and give up the remainder to it. As with an anode, the associated power is made up of two parts, i.e. the thermal power acquired in the column and that due to the acceleration experienced by the ions in the cathode-fall region, equal to the cathode-fall voltage and the ionic currents.

(b) The power given up when ions and electrons recombine at a cathode is equal to the product of the ionisation potential of the atoms, in volts, and the ion current flowing into the cathode.

(c) Power is also given to a cathode in other ways, as outlined in Section 3.3.1.1 (c) and (d), and again these quantities tend to be small.

3.3.1.4   Power lost by a cathode

In addition to dissipating power by the mechanisms described in Section 3.3.1.2 for an anode, a cathode supplies significant power to the electrons emitted by it, this quantity being equal to the product of the electronic current and the work function (a voltage).

3.3.1.5 Power balance

The difference between the power supplied to and the power dissipated from a pair of electrodes is used in raising their temperatures and in melting and vaporising them. The power available for this purpose at an anode was given by Cobine and Burger [24] as:

$$\text{power} = (V_{\text{af}} + V_{\text{wf}} + V_{\text{T}}) \times i \tag{3.2}$$

Since the ratio of electronic to ionic currents at a cathode is unknown it is not possible readily to determine the power supplied to it. X-ray photographs of elements which have passed high currents invariably show that associated pairs of anodes and cathodes have burned back almost equally and therefore it is reasonable to assume that the power supplied to a cathode is the same as that given to an anode.

An important factor, which affects the burnback, is that only a fraction of the element material which initially occupies the space in which a gap is ultimately formed is vaporised during the arcing process, the remainder flowing away in liquid form. This fact was revealed by Wright and Beaumont when elements which had cleared large currents were examined microscopically. The elements together with the fused quartz filler, known as fulgurite, were sectioned longitudinally and in a plane orthogonal to that of the element strip, for examination. An example is shown in Figure 3.4 and the metal which flowed out in liquid form can be seen clearly. Further tests showed that the metal which had vaporised was deposited finely in the quartz filling material. This behaviour is presumably caused by the pressure distribution during arcing.

In calculating the arcing performance by taking account of energy requirements and interchanges, it is clearly necessary to determine the proportion of the element which is vaporised as this requires more energy per unit mass than the material which runs away in liquid form. To assess the proportions, Wright and Beaumont used a number of fuselinks containing silver elements and quartz filler to clear large currents. They were subsequently X-rayed to determine the final lengths of the arcs. The time integrals of the currents during the arcing periods were determined from oscillograph records of the currents and the energy available for melting and vaporising was taken to be:

$$E_{\text{MV}} = 2(V_{\text{af}} + V_{\text{wf}} + V_{\text{T}}) \int_0^{t_a} i \, dt \tag{3.3}$$

in which t_a is the duration of the arcing period.

This was equated to the energy needed to remove the electrode material, i.e.

$$E_{\text{MV}} = m_{\text{v}}(\text{latent heat of vaporisation}) + m_{\text{t}}(\text{latent heat of fusion})$$
$$+ m_{\text{t}}(\text{melting temperature} - 200) \times \text{specific heat} \tag{3.4}$$

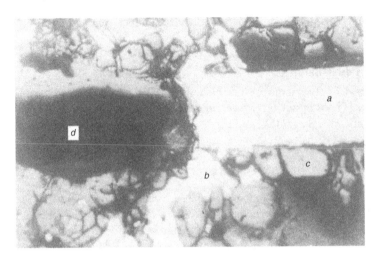

Figure 3.4   Microscopic picture of fulgurite section

Where
a   electrode
b   electrode material which has flowed away in liquid form
c   grains of filler
d   arc cavity which has been filled with epoxy adhesive to enable the
     section to be made through the fulgurite

in which m_t is the total mass of material removed from the element, a quantity determinable from the X-ray photographs, and m_v is the mass of the vaporised metal.

It will be noted that this latter expression neglects the energy required to raise the temperature of the element material from the melting to boiling temperature, a quantity which is very small compared with the latent heat of vaporisation. It will also be seen that the bulk temperature of the electrode material is assumed to be raised to 200°C during the pre-arcing period.

In all the tests which were done it was found that eqns. 3.3 and 3.4 were satisfied when $m_v \simeq 0.40\, m_t$, i.e. the mass of material vaporised was only about 40 per cent of the total mass which was melted. For this condition, it is found that 20 per cent of the total energy supplied is used to provide the latent heat of fusion L_f and consequently the following expression may be used to determine the lengthening of the gap between a pair of electrodes, that is the burnback in an interval of time δt:

$$l_2 - l_1 = \frac{\text{mass of electrode removed in } \delta t}{A_e \times \text{density of electrode material}}$$

$$= \frac{2(V_{af} + V_{wf} + V_T) \times \{(i_1 + i_2)/2\}\delta t \times (0.2/L_f)}{A_e \times \text{density of electrode material}} \tag{3.5}$$

where l_1 and l_2 and i_1 and i_2 are the gap lengths and currents at the beginning and end of the interval, respectively, and A_e is the cross-sectional area of the electrode at the arc roots over the time being considered.

By continuously calculating the burnback for each time interval the length of the positive column at any instant in the arcing period can readily be determined.

3.3.2 Cross-sectional area of a positive column

When a positive column is established in a fuse it receives power from the electrical system equal to the product of the voltage drop along the column and the total current in it. In any interval of time δt the energy given to the column may be expressed as:

$$\text{energy input } (E_a) = \tfrac{1}{2}(v_{a1}i_1 + v_{a2}i_2)\delta t \tag{3.6}$$

in which v_{a1} and v_{a2} are the column voltages and i_1 and i_2 are the currents at the beginning and end of the interval, respectively.

During the earlier part of the arcing period some of the energy input is retained within the column and is responsible for its increases in dimensions and temperature, while in the later part of the period the column reduces in cross-sectional area and the energy within it decreases. The changes in the energy present in a column during a short period may, however, be shown to be very small compared with the total energy input and so also may the energy given up to the electrodes. It may therefore be assumed that the power given to a column at any instant is dissipated to the surrounding filler which will consequently melt back progressively. About 2100 J are required to produce a gram of molten quartz and thus about $E_a/2100$ g of molten quartz [25] must be produced in the time interval δt.

Fuselinks are vibrated after the filling material is put into them and this should achieve random packing, in which state about 60 per cent of each unit of volume within the body should be occupied by quartz, the remainder being taken up by air. On melting, quartz increases in volume by about 7 per cent but the grains fuse together and as a result the volume of the liquid quartz is only about 64 per cent (i.e. $1{\cdot}07 \times 60$) of that originally occupied by the quartz–air mixture. A separate air space or cavity is formed into which the column can expand. In addition, a column expands into the space previously occupied by the part of the element which has melted. The total increase in the volume of a column in an interval δt may therefore be taken to be

$$\text{vol}_{a2} - \text{vol}_{a1} = \left\{ \frac{(1 - 0{\cdot}64)}{2 \times 2100 \times 1{\cdot}6 \times 10^6}(v_{a1}i_1 + v_{a2}i_2)\delta t \right.$$

$$\left. + 2(l_2 - l_1)A_e \right\} \text{ cubic metres} \tag{3.7}$$

in which l_1 and l_2 are the column lengths at the beginning and end of the interval, respectively, and A_e is the element area at the arc roots. It should be noted that the above expression uses a value of $1{\cdot}6 \times 10^6$ g/m^2 for the density of dry, randomly packed sand.

Examinations of fuses which have cleared large currents have confirmed that there is always a cavity in the fused filler, surrounding each of the positions at which there had been a restriction in the fuse element. Although the cavities do not have a constant cross-sectional area it may be thought reasonable for mathematical purposes to neglect this effect and to take the column area at the end of the interval as being

$$A_{a2} = \frac{vol_{a2}}{l_2} \tag{3.8}$$

3.3.3   Electrical conductivity of a positive column

When a positive column is established, material evaporated from the electrodes often enters it in the form of jets. Wright and Beaumont conducted a number of tests on fuselinks, including the taking of high-speed films of the arcs, and it was established that jets were present. The jets from opposite electrodes collide and create turbulence which in turn causes the temperatures and electron densities within each column to be relatively uniform.

In any short interval of time δt, a number of atoms of the vaporised electrode material are accelerated in jets into the column and a fraction X of them become ionised. During the same interval an almost equal number of atoms and ions must be scattered out of the column, as otherwise untenable pressures would be set up and the conductivity would be much above the values it can be shown to have by analysing voltage and current traces obtained during the testing of fuselinks. For the purposes of modelling, it may be possible to assume that the following equations apply for any time interval:

$$N_{ai} = N_a \tag{3.9}$$

and

$$N_e = X N_a \tag{3.10}$$

in which N_{ai} and N_e are the numbers of atoms and ions and electrons, respectively, scattered out of the arc in δt seconds and N_a is the number of atoms evaporated from the electrodes in the same period.

It was stated in Section 3.3.1.5 that the total mass of electrode material which is melted in a time interval is given by

$$\text{mass melted} = 2(V_{af} + V_{wf} + V_T)\left(\frac{i_1 + i_2}{2}\right)\left(\frac{0.2\delta t}{L_f}\right)\ \text{grams}$$

Of this amount 40 per cent is vaporised, and therefore knowing the number of atoms per gram (N_g) of the element material (for silver $N_g = 5.66 \times 10^{21}$) it is clear that the number of atoms injected into the jets in an interval δt is

$$N_a = N_g m_v$$

A positive column loses energy in the following main ways:

(a)   The kinetic energy of the atoms and ions (N_{ai}) which are scattered out of the column is given to the surrounding filler. The amount given up in an interval δt is

$$KE_{ai} = 1{\cdot}5 N_{ai} k_B \frac{(\theta_{a1} + \theta_{a2})}{2} \tag{3.11}$$

where k_B is Boltzmann's constant which is equal to $1{\cdot}38 \times 10^{-23}$ J/K, and θ_{a1} and θ_{a2} are the column temperatures at the beginning and end of the interval, respectively.

(b) The kinetic energy of the electrons (N_e) scattered out of the column is also given to the filler. The amount given up in an interval δt is

$$KE_e = 1{\cdot}5 N_e k_B \frac{(\theta_{a1} + \theta_{a2})}{2} \tag{3.12}$$

(c) The energy required to ionise N_e atoms is taken out of the column in a time δt and given up to the filler material when recombination occurs. The energy is given by

$$IE = N_e E_J \tag{3.13}$$

in which E_J is the ionisation energy per atom, the value of which is 12.1×10^{-19} J for silver.

(d) Energy is lost by radiation to the surroundings of the column. Because the column temperature and pressure are both high, the emissivity may reasonably be assumed to be unity, i.e. black-body radiation. To further simplify the calculation, the temperature difference may be taken to be the temperature of the column because of the relatively low temperature of its surroundings. With these approximations the radiation energy loss may be expressed as:

$$RE = \text{surface area of column} \times k_s \left(\frac{\theta_{a1} + \theta_{a2}}{2} \right)^4 \tag{3.14}$$

where k_s is the Stefan–Boltzmann constant which is equal to $5{\cdot}67 \times 10^{-8}$ W/(m^2 K^4).

For modelling purposes, Wright and Beaumont assumed the column to be cylindrical and its diameter was taken to be

$$d_{a2} = \left(\frac{4}{\pi} A_{a2} \right)^{0.5}$$

in which A_{a2} is the cross-sectional area of the column at the end of the time interval. The surface area of the column is then

$$\text{surface area} = \pi d_{a2} l_2 = 2l_2 (\pi A_{a2})^{0.5} \tag{3.15}$$

Energy is also lost from a column in other ways including conduction to the filler and radiation to the electrodes, but the amounts may be shown to be relatively small and the thermal time constants are high.

The energy which enters a column is the time integral of the product of the voltage along it and the current flowing through it. For a short interval of time this may be expressed as

$$\text{input energy} = \frac{V_{a1} i_1 + V_{a2} i_2}{2} \delta t$$

The following energy-balance equation must apply:

$$\frac{V_{a1}i_1 + V_{a2}i_2}{2}\delta t = KE_{ai} + KE_e + IE + RE$$

Substitution in this equation from eqns. 3.9–3.15 gives

$$\frac{V_{a1}i_1 + V_{a2}i_2}{2}\delta t = N_a\left[1{\cdot}5k_B\left(\frac{\theta_{a1} + \theta_{a2}}{2}\right)\left\{1 + \left(\frac{X_1 + X_2}{2}\right)\right\} + \frac{X_1 + X_2}{2}E_J\right]$$
$$+ 2\left(\frac{l_1 + l_2}{2}\right)k_s\left(\frac{\theta_{a1} + \theta_{a2}}{2}\right)^4\left(\pi\frac{A_{a1} + A_{a2}}{2}\right)^{0{\cdot}5} \tag{3.16}$$

At any time, the temperatures of the atoms and electrons in a column will be the same because the mean free path is very short and they must experience many collisions before leaving the column. In these circumstances Saha's equation [26], which relates electron density, ionisation fraction and temperature is applicable, i.e.

$$n_{e2} = \frac{1 - X_2}{X_2}\text{antilog}_{10}\left(\frac{-3{\cdot}878 \times 10^4}{\theta_{a2}} + 1{\cdot}5\log_{10}\theta_{a2} + 15{\cdot}385\right) \tag{3.17}$$

in which n_{e2} is the electron density at the end of the interval. The atoms evaporated in an interval δt flow in the vapour jets which may, for modelling purposes, be assumed to be of the cross-sectional areas of their associated electrodes and to have a constant velocity throughout the column.

The volume (V_a) occupied by the atoms is given by

$$V_a = A_a V_J \delta t$$

in which V_J is the velocity of the jets.

The atomic density of this material is therefore

$$n_a = \frac{N_a}{A_e V_J \delta t}$$

and the corresponding electron density is

$$n_e = X N_a = \frac{N_a X}{A_e V_J \delta t} \tag{3.18}$$

Wright and Beaumont further assumed that the ions and electrons reach terminal velocities in a column before being scattered out of it, and this enabled ionic and electronic conductivities to be assigned. Because the electrons move at much higher velocities than the ions, a further simplification of neglecting the ionic component of the current may be made.

At ionisation levels of more than 0·01 per cent which is certainly the situation in fuse arcs, Spitzer's conductivity equation [27] given below is valid:

$$\sigma_2 = \frac{1{\cdot}55 \times 10^{-2}\theta_{a2}^{3/2}}{\log_e\{(1{\cdot}242 \times 10^4\theta_{a2}^{3/2})/n_{e2}^{0{\cdot}5}\}} \tag{3.19}$$

Eqns. 3.16–3.19 are non-linear and as they relate five unknown terms, i.e. $\sigma_2, n_{e2}, \theta_{a2}, V_J$ and X_2 they cannot be solved.

Wright and Beaumont performed a range of tests on fuselinks, and the variations of the electrical conductivities were determined from the current and voltage traces. These values were then substituted in eqns. 3.16–3.19 and the other four unknown quantities were calculated. It was found that the jet velocity V_J varied over a considerable range, which did, however, correspond well with values determined by other workers [27].

Because the variations in jet velocity did not have a great effect on the column conductivity, Wright and Beaumont used a value of vapour-jet velocity in the above equations, which minimised the errors between the computed and measured values of conductivity for the tests which had been done.

3.4   Complete mathematical model

As stated earlier it is usually possible to regard a circuit as having a source of EMF and an associated network of series and shunt impedances. In many cases the shunt branches do not have a significant effect, particularly during fault conditions, and the circuit may often be represented simply as a source of EMF e_s in series with a resistance and an inductance (R_c and L_c, respectively). In these circumstances, the general non-linear equation which would apply during the arcing period would be

$$c_s = i(R_a n + R_c) + \frac{d(L_c i)}{dt} + n(V_{af} + V_{cf}) \tag{3.20}$$

in which R_a is the resistance of each arc column, V_{af} and V_{cf} are the anode- and cathode-fall voltages, respectively, and n is the number of arcs or restrictions in the fuse element. When step by step computation is to be used to determine the current variation from the end of the pre-arcing period, it is convenient to rewrite eqn. 3.20 in the form

$$\frac{e_{s1} + e_{s2}}{2} = \frac{n(i_1 R_{a1} + i_2 R_{a2})}{2} + \frac{i_1 R_{c1} + i_2 R_{c2}}{2} + \frac{L_{c1} + L_{c2}}{2} \times \frac{i_2 - i_1}{\delta t}$$
$$+ \frac{i_1 + i_2}{2} \times \frac{L_{c2} - L_{c1}}{\delta t} + n(V_{af} + V_{cf}) \tag{3.21}$$

The resistance of the column at the end of any step is given by

$$R_{a2} = \frac{v_{a2}}{i_2} \tag{3.22}$$

and also by

$$R_{a2} = \frac{l_2}{\sigma_2 A_{a2}} \tag{3.23}$$

It must be recognised that eqns. 3.5, 3.7, 3.8, 3.16–3.19 and 3.21–3.23 are independent and include ten unknowns, $i_2, l_2, vol_2, A_{a2}, \sigma_2, n_{e2}, \theta_{a2}, X_2, R_{a2}$ and v_{a2}. They can therefore be solved if $i_1, l_1, vol_1, A_{a1}, \sigma_1, n_{e1}, \theta_{a1}, X_1, R_{a1}$ and v_{a1} are known.

For the first time interval, the values of e_{s1} and i_1 must be those which are obtained at the end of the pre-arcing period and the parameters l_1, vol_1, A_{a1}, θ_{a1} and X_1 at the beginning of the interval are very small and may therefore be taken to be zero. The only remaining parameter which needs to be known is the voltage along the column. Oscillograms obtained during many varied tests have shown that the total voltage across each notch of a fuselink element changes abruptly from a small value during the pre-arcing period to about 50 V at the instant of arc initiation. The exact value depends on the dimensions of the restriction, long restrictions having slightly higher initial voltages than short ones. For modelling purposes, the initial value of the voltage across the column was taken by Beaumont and Wright to be 33 V.

Step-by-step calculations of the type described above should enable the characteristics during the arcing period to be determined.

Many parameters associated with fuse arcs such as pressure, temperature and length have been calculated using the above model. As an example, the variation of arc temperature with time in a fuselink, for which the current and voltage waveforms were as shown in Figure 3.5, is shown in Figure 3.6. It can be seen that the temperature within the arc varies within the range 10×10^3 to 15×10^3 K.

At the time these theoretical studies were undertaken, it was not possible to compare the various results with values obtained by experiment because the transducers then available could not be inserted into fuselinks without significantly affecting their performance. Subsequently, however, Barrow and Howe [28,29] conducted laboratory tests on fuselinks into which optical fibres were inserted, the fibres being of similar material to the filling material in the fuselinks, so that they did not appreciably affect the performance.

Initially a row of optical fibres was inserted in each fuselink as described in Reference 28 and shown in Figure 3.7, the inner ends of the fibres being near one face of the element. As a fuselink operated when a high current passed through it, light

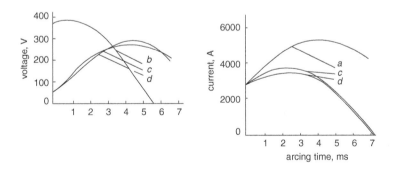

Figure 3.5   *Current and voltage waveforms*

Where
a   waveform of prospective current
b   waveform of open-circuit voltage
c   computed waveforms
d   waveforms obtained by experiment

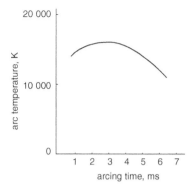

Figure 3.6   Variation of arc temperature with time

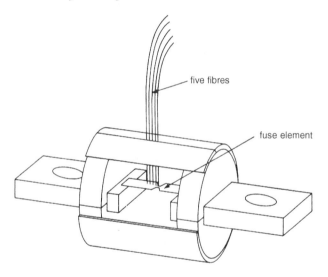

Figure 3.7   Fuselink containing optical fibre

emanating from the arc was transmitted along the fibres and detected and recorded by external equipment. Light was received first from the fibre positioned at the centre of the restriction in the element and then light was received in sequence from the other fibres as the arc extended. It was therefore possible to determine the rate of burnback, i.e. the rate at which the arc extended, and comparisons with values determined experimentally could be made.

In a second series of experiments, described in Reference 29, the light emissions from optical fibres inserted into fuselinks were determined during arcing periods. A rapid-scanning spectrometer and photomultiplier tube were used to measure the relative intensities of the most prominent wavelengths in the emitted light, and this enabled the arc temperatures to be calculated and compared with theoretically determined values.

Subsequent related experimental and theoretical studies conducted by Cheim and Howe are described in References 30 and 31. It is believed that continuation of such work will enable refined and accurate modelling of fuse behaviour to be achieved and that it will eventually be possible to determine the effects of varying individual parameters, such as the specific heat of the filling material or the amount by which it expands on liquifying. In this way the most suitable materials and optimum dimensions of a fuselink for a particular application could be determined. Its performance for a wide range of system conditions including prospective current, instant of fault occurrence in the voltage cycle and fault-circuit power factor could also be computed and then only a small number of experimental checks should be needed to provide confirmation. In this way it should be possible, in the future, to reduce greatly the amount of testing from that needed at present, and significant financial savings may consequently be effected.

Recent work by Rochette *et al.* [32] introduced a model describing the mechanical interaction and energy transfer mechanisms from the arc column to the porous granulated surrounding medium. The heat and mass transfers of the solid, liquid and vapour phases were modelled using Darcy's and Forchheimer's laws. The results produced simulated fulgurites with characteristics similar to those observed in practice.

The preceding arcing models are only useful for the interruption of high short-circuit currents, for which multiple arcing occurs in the notch zones, and arc extinction occurs when the limited current reaches zero. In this case the resistive shunting effect of the hot fulgurite minimises the possibility of restriking [33].

For the interruption of low overcurrents the situation is different. Because the heating rate is much slower, the cooling effect of the ends has great effect, and the central region of the fuse element becomes much hotter than its surroundings. A notch in the central part of the fuse melts first, producing a single arc with a relatively low voltage because the current is low. There is little current-limiting effect at first, but the arc voltage increases as the element burns back. When the first current zero is reached, the arc extinguishes, but the gap left may be too short; the arc restrikes, and a second half cycle of arcing begins. This may continue for several half-cycles, until the gap produced by burnback is long enough to withstand the restriking voltage. This process has been studied by Erhard *et al.* [34], who obtained extensive test data and developed a model which included the effects of sequential burnback and restriking.

Chapter 4

Constructions and types of low-voltage fuses

It was stated in Chapter 1 that fuses are classified into three groups, namely low-voltage, high-voltage and miniature. This chapter will deal with the construction of fuses in the first category and the other categories will be described in Chapters 5 and 6. For clarity in each of the chapters, British practice will be described first and then the practices in other parts of the world will follow.

This chapter will deal in a general way with the constructions of low-voltage fuses produced for industrial and domestic applications and also those suitable for the protection of semiconductor devices.

British low-voltage fuses may be divided into two groups, both of which are very widely used. The first group employs cartridge fuselinks which are for use in power frequency AC circuits at rated voltages not exceeding 1000 V and DC circuits not exceeding 1500 V. They are produced with rated currents generally in the range 2–1250 A. They are often mounted in fuse carriers which are mounted in turn in fuse bases, a particular example being shown in Figure 4.1. The second group, which is of diminishing importance, uses directly replaceable elements, of wire form, which are mounted in suitable fuse carriers and bases. These are officially described as semi-enclosed fuses but they are commonly referred to as rewireable fuses. They are produced with rated currents up to 100 A and are suitable for use in AC circuits with voltages up to 240 V to earth. A fuse of this type is shown in Figure 4.2. Details of the constructions of both types are given below.

4.1 Cartridge fuses

The fuselink is the component which must be designed to provide the necessary electrical operating characteristics, and in this sense its construction requires the greatest study.

Figure 4.1   Typical fuse carrier and base

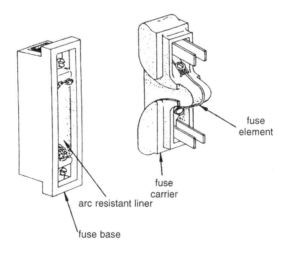

Figure 4.2   Semi-enclosed fuse

Assembled and exploded views of a typical fuselink are shown in Figure 4.3. The various parts will be considered separately next and then sections dealing with fuse carriers and bases will follow.

4.1.1   Fuselink elements

Cartridge fuselinks, as stated earlier, contain one or more parallel-connected elements which are made of materials of low resistivity. The materials, if possible, should also

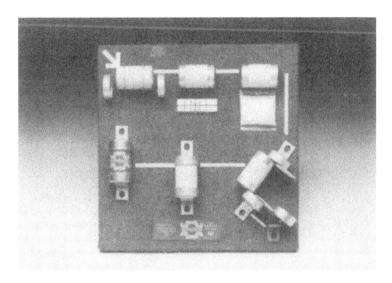

Figure 4.3   Assembled and exploded views of cartridge fuselinks

possess the following properties if rapid operation is required:

(*a*) low specific heat
(*b*) high thermal conductivity
(*c*) low melting and vaporisation temperatures
(*d*) low latent heats
(*e*) low density
(*f*) ease of connection to other conductors.

It will be appreciated that rapid operation is usually required during short-circuit conditions. Taking a wire element, of length l and cross-sectional area A, and for simplicity considering operating times short enough for heat movement within the fuselink to be neglected, it is clear that the following energy balance must be satisfied at the instant when vaporisation of the element occurs:

$$\frac{1}{A} \int_0^{t_v} i^2 \rho \, dt = Al \times \text{element density } \{\text{specific heat}(\theta_v - \theta_i)$$
$$+ \text{latent heat of fusion}\}$$

in which θ_v and θ_i are the vaporisation and initial temperatures, respectively, and ρ is the resistivity of the element.

The time taken (t_v) for vaporisation to occur is therefore proportional to the square of the cross-sectional area multiplied by the element density [specific heat $(\theta)_v - \theta_i$) + latent heat of fusion] and it varies somewhat inversely with the resistivity. This tends to confirm that the time to vaporising and the energy let-through to the protected circuit ($\int_0^{t_v} i^2 \, dt$) are both minimised when materials with the desirable properties (a) and (c), listed above, are employed. Short operating times at high currents are not the only requirement, particular operating time/current characteristics being necessary to enable fuselinks to fully protect the devices or circuits with which they are associated. In addition it is usually desirable that they should have a reasonably low fusing factor, that is the ratio of the minimum fusing current to the rated value should not be greatly in excess of unity, so that adequate protection is given against prolonged overloads. The term minimum fusing current is no longer used in fuse standards, despite its descriptive nature. For general purpose fuselinks, type gG to *IEC 60269-1* (*BS88-1*) fuses are subjected to a conventional non-fusing current, I_{nf}, and a conventional fusing current, I_f, for a time equal to the conventional time. The conventional time relates to the thermal time constant of the whole fuse and can vary from 1 to 4 h. Typical values for I_{nf} and I_f are:

$$I_{nf} = 1.25I_n$$
$$I_f = 1.6I_n$$

In addition, a cable overload test is made at $1.45I_z$, I_z being the associated cable rating, see Section 7.3.

At the rated current level and currents up to the minimum fusing level, steady-state equilibrium conditions must be established in which the electrical power input to an element is conducted away from it and dissipated from the fuselink outer surfaces. This is the reason for the property (b) listed above. Of course, not only the element must have a high thermal conductivity but so must the surrounding filling material, the body and other parts of the fuselink. In addition, because the power which can be dissipated from a fuselink of particular dimensions, with surface temperatures limited to acceptable levels, is fixed, it is necessary to limit the power produced by the element. This can only be achieved by having an element with a relatively large cross-sectional area and low resistivity. It will be seen that these requirements conflict with those which were shown to be needed to give rapid clearance of short circuits. Practical conditions, including cost, lead to acceptance of small cross-sectional areas of low-resistivity material.

It will further be appreciated that it is difficult to limit the temperature rises of fuselink surfaces to low levels at currents approaching the minimum fusing level if the element material has a high melting-point temperature and this is a further reason for attempting to satisfy property (c) above. A low fusing factor is very difficult to obtain when plain elements of materials with high melting points are employed, because the element temperature at rated current must be high and this must tend to lead to unacceptably high fuselink surface temperatures. Fortunately, acceptable fusing factors can be obtained when using elements with high melting points by alloying them to other metals to take advantage of the M-effect described in Section 2.3.

After extensive studies of all the available materials and taking the above factors into account, it has been concluded that silver and copper are the most suitable materials and certainly they are used in the vast majority of modern fuselinks.

Strip-type elements with restricted sections are produced in various forms by different manufacturers, a few typical examples being shown in Figure 4.4. The dimensions and numbers of restrictions depend on the current and voltage ratings. The strip thicknesses are usually in the range 0·05–0·5 mm. Fuselinks with

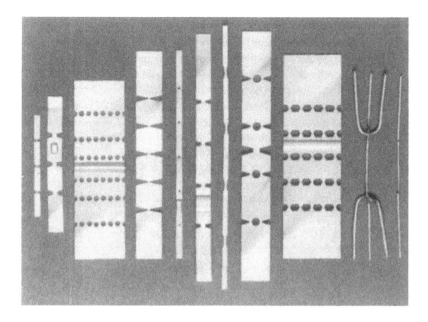

Figure 4.4   Various fuse element designs

single-strip elements are used for rated currents in the range of about 10–63 A and fuselinks with higher ratings contain two or more elements connected in parallel.

Although fuselinks containing several wire elements connected in parallel were produced in the past, this practice has now ceased and only single-element designs are produced for industrial fuselinks using wires up to approximately 0·2 mm diameter for ratings up to 10 A. It is of interest to note that ratings as low as 0·25 A can be produced.

4.1.2   Fuselink bodies

The element or elements are welded or soldered to plated copper or brass inner end caps, which together with the body form an enclosure or cartridge. The ends caps are firmly fixed to the fuse body and usually an interference fit.

The bodies must possess good electrical insulating properties and should not allow the ingress of moisture. They should be reasonably good thermal conductors and have an adequate emissivity constant so that they can transmit and radiate heat energy being dissipated by the elements within them and, in addition, they must be mechanically robust and capable of withstanding thermal shock during operation. Ceramic is used almost exclusively for this purpose and the bodies are invariably cylindrical. Such bodies have wall thickness in the range 2–15 mm and outer diameters up to about 100 mm.

4.1.3   Filling material

Cartridge fuselinks which are required to have high breaking capacity are invariably filled with granular quartz of high chemical purity and grain sizes in the region of 300 μm. The grain size is tailored to suit the element thickness and desired performance.

The filling material conducts some of the heat energy away from the fuse element to the body and therefore to obtain consistent performance it is necessary that the packing density of the filling material is maintained constant during production. This factor will have a very significant effect on the behaviour at high current levels because a low packing density may allow the arcs to expand more rapidly, affecting the column voltage and thus the rate of current change.

To ensure uniformity of packing, the quartz is poured into the open ends of fuselinks which are complete except for the pressing on of their final outer end cap and tag assemblies. The fuselinks are vibrated and, if necessary, extra quartz is added during this process to ensure that the internal space is packed as densely as possible. Each fuselink assembly is then completed by adding the final outer end cap and tag assembly.

4.1.4   Industrial fuses

The majority of industrial applications are catered for by fuselinks with dimensions and performances specified in *IEC 60269-2-1*. *Section II* of this document covers fuselinks with terminations which enable them to be bolted into their circuits thus ensuring that good connections are attained. This is clearly important in applications where the rated currents of the circuits are high. *Section IV* of *IEC 60269-2-1* (*BS88-6*) covers fuselinks with blade terminations which allow them to be fitted directly into spring contacts. This arrangement facilitates the replacement of fuselinks.

IEC 60269-2-1 Section II includes fuselinks from 2 to 1250 A. These fuselinks are fitted with end caps that have integral spade-type terminations with holes or open-ended slots in them. Typical examples of both the above types of fuselinks are shown in Figure 4.5. Cartridge fuselinks with current ratings above about 100 A are usually installed directly in convenient positions without their own special or standardised housings. In addition, they are incorporated in fuse-combination switches, an illustration being shown in Figure 4.6.

Fuselinks with rated currents up to 100 A are usually fitted into fully shrouded fuse-holders comprising a carrier and a base. These are used in large numbers in distribution fuse boards and switch-disconnector-fuses; typical examples are shown in Figures 4.7 and 4.8. In addition many individual fuses are used in a wide range of applications.

An illustration of a modern fuse-holder has already been referred to (Figure 4.1). It shows the cable entries at the ends of the fuse base. This, the standardised method is termed front connection or front entry.

The present-day fuse-holder has many safety features, one being that live metal cannot be touched when the fuse carrier is being removed from or inserted into the

Figure 4.5   Tag-type fuselinks

Figure 4.6   Fuse-combination-unit

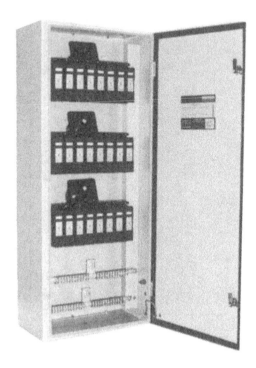

Figure 4.7   Distribution board

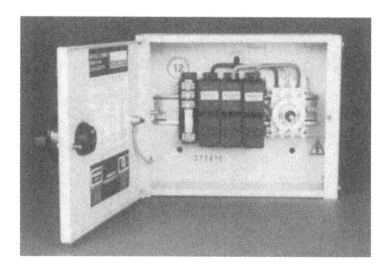

Figure 4.8   Switch-disconnector-fuse

Figure 4.9   Lockable safety carrier

fuse base and a second is that the fuse base must be fitted with barriers or shutters so that live metal is exposed within it when the fuse carrier is removed.

As an additional precaution, lockable safety-fuse carriers of the form shown in Figure 4.9 are now produced. They do not contain fuselinks and they may be inserted into fuse bases and locked when normal fuselinks and holders are removed prior to maintenance work being undertaken. This action clearly prevents a circuit being re-energised whilst work is still in progress.

As stated earlier, *IEC 60269-2-1 Section IV* (*BS 88-6*) covers compact fuses incorporating fuselinks with blade terminations. An example is shown in Figure 4.10.

The compactness of these fuses and the ease with which the fuselinks can be replaced has made them very popular. They are now widely used to provide the over-current protection of final distribution circuits, being incorporated in both switch-fuses and fuse boards. A novel fuse distribution board suitable for circut ratings up to 63 A at 415 V, which is adequate for most applications, is shown in Figure 4.11. It incorporates modular fuse-holders, rated at 63, 32 and 16 A, which slot into rising

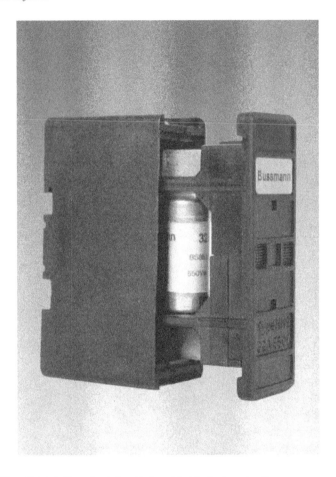

Figure 4.10   British Standard fuselinks with blade terminations

shrouded 200 A busbars. An integral isolator may also be incorporated if required by the customer. This is illustrated in Figure 4.11 together with a traditional board.

Some large organisations, in the past, produced specifications and standards to meet their own special needs, two of these being the UK Electricity Supply Industry and Ministry of Defence.

From its early days, the UK Electricity Supply Industry has used distribution fuses not only to provide protection but also to act as load-breaking switches and isolators. An early type of fuse used for this purpose is illustrated in Figure 4.12. These were sometimes called jug-handle fuses, and they were available in current ratings up to 500 A. Not surprisingly, instructions were issued that an operator must close his or her eyes and shelter his or her face with an elbow when reclosing these fuses but, as an alternative, a 1·22 m (4 ft) wooden stick with a clip on the end could be used to manipulate them. This basic design was improved, but it was not until the mid-1930s

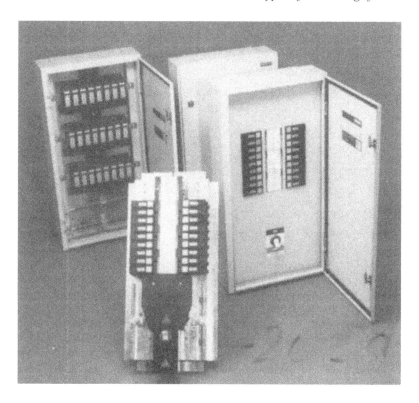

Figure 4.11   Modern distribution board

that fuses with a wedge-action device for tightening the fuselinks to the base were employed.

Standardised cartridge fuselinks with slotted tags having basic fixing centres of 82 and 92 mm, capable of carrying maximum rated currents of 400 and 800 A, respectively, are used. They are produced to meet the requirements of *BS 88-5* and the recently introduced *IEC 60269-2-1 Section VI*.

In accordance with *IEC 60269-2-1 Section VI*, fuselinks with wedge tightening contacts have a characteristic designated 'gU'. Compared with the general characteristic 'gG' to *IEC 60269-1*, this is faster acting at higher fault currents and thereby improves discrimination with upstream MV protection in most electricity distribution networks. A typical fuse together with the type of wedge tightening contacts used is shown in Figure 4.13.

The above fuses were extensively used in open type substation boards, underground disconnection boxes, distribution pillars and heavy duty service cut-outs. However, as requirements for operator independent safety increase, these earlier designs that rely on the removal of a fuse handle that has no proven switching capability to isolate a circuit and which exposes operators to live conductors, are increasingly unacceptable. Where such early designs are employed, their operation relies on the

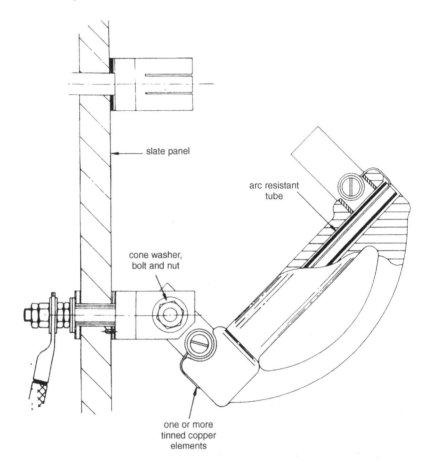

slate panel

arc resistant
tube

cone washer,
bolt and nut

one or more
tinned copper
elements

Figure 4.12   Early distribution fuse

use of suitably skilled operators wearing personnel protective equipment, including gloves and visor.

In order to be more in line with present safety expectations, new installations are generally provided with as a minimum, shielded fusegear. These forms of equipment utilise fuse handles as shown in Figure 4.14, and have all live conductors shrouded so as to provide operators and any one in the vicinity of the equipment, with protection against direct contact in the normal service condition (all fuse handles in place). Switching of circuits on shielded fusegear does however require the same level of skill and safety precautions as the earlier designs of fusegear.

For network managers who wish to retain the benefits of single-phase operation and ensure safety independent of the skill of the operator, switched and insulated fusegear (SAIF) is the preferred solution. These designs of equipment include the standard fuselinks designed for wedge tightening contacts and provide protection for operators against direct contact with live conductors in the normal service condition,

Figure 4.13   Fuse incorporating wedge action

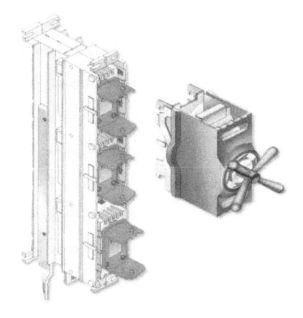

Figure 4.14   Shielded fusegear

while switching circuits and when changing fuselinks. To switch a circuit, SAIF employs a portable mechanism that latches onto the fuse base. This, in turn, uses stored spring energy to move the fuse carrier between open and closed positions in a switching action. Switching performance, fault make, load break, is proven by test,

Figure 4.15   An SAIF unit

identical to that of a conventional fuse switch disconnector with independent manual operation and in accordance with *IEC 60947-3*. Figure 4.15 shows an SAIF unit.

The UK Ministry of Defence produced specifications *DEF63A* and *DEF64* in 1961 and 1964, respectively, to cover a completely standardised range of fuselinks and fuse-holders to meet the needs of all the armed services. They were suitable for use at voltages up to 440 V AC and 230 V DC and were produced with current ratings up to 800 A, but mainly used up to about 100 A. Some of these are shown in Figure 4.16. Very rigorous type tests were required to prove the ability of the fuses to withstand the worst environmental conditions likely to be encountered in service. Included were tests to prove that the fuses would not be damaged by vibration, acceleration, bumps, shocks, extreme climatic conditions, tropical exposure and mould growth.

The specifications included standardisation of the elements and this, together with the compact dimensions, created problems in developing the fuse system to meet modern requirements, particularly with regard to very high breaking capacity. As a consequence, the Ministry updated its standards in 1980 and 1981. The requirements for fuselinks are covered by *Defence Standard 59-96* and the fuse holders by *Defence Standard 59-100*. These standards cover two types of fuses: first the preferred types for new equipment which must satisfy the requirements of *BS 88-2* and also the Ministry of Defence environmental type approval tests; secondly the types originally produced to *DEF63A* and *DEF64* are covered and retained for replacement purposes. Both types must be subjected to quality-assurance procedures approved by the Ministry.

Figure 4.16   Ministry of Defence fuses

4.1.5 Domestic fuses

These are defined as being for use in domestic and similar premises, e.g. dwelling houses, blocks of flats and office buildings. Cartridge fuselinks are produced for use in plugs, consumer units and electricity-supply-authority fuses. In view of the similarity in construction of plug fuselinks with miniature fuselinks, these are dealt with in Chapter 6. The consumer-unit and supply-authority fuselinks are covered by *BS 1361* and categorised as Type I and Type II, respectively.

These fuselinks are fitted with plain end caps without projecting terminations. These are called ferrule-type fuselinks and examples of them are shown in Figure 4.17.

Type I fuselinks are produced in four non-interchangeable sizes and their main application is in situations where high fault levels arise, their rated breaking capacity being 16·5 kA.

Type II fuselinks are those belonging to the supply authority which are fitted in domestic premises. They are often called 'house-service' fuselinks.

Various forms of rewireable fuse were used in the early days, and even in the late 1940s semi-enclosed fuses housed in iron boxes with hinged covers were still being installed, although bakelite boxes had been available since about 1935. From 1940 onwards, cartridge fuses in bakelite boxes began to supersede the rewireable fuses, but it took some time for the advantages of their consistency and high breaking capacity to be recognised sufficiently for their use to become widespread.

A cartridge fuselink, rated up to 60 A, was produced to meet electricity-supply-industry requirements, but after some time the increasing use of electricity in the home made it necessary to introduce a larger design rated up to 100 A. These two sizes, now regarded as the standards, are referred to as Types IIa and IIb in *BS 1361*,

Figure 4.17   Ferrule fuselinks

which gives the performance requirements with which they must comply. The power loss of the fuselinks is limited and the permitted loss from the 100 A size is only 6 W.

The requirements with which the complete supply-authority fuse unit must comply are specified in *BS 7657* formerly Electricity Supply Industry (*ESI*) *Standard 12-10*. A typical unit with fuselink is shown in Figure 4.18.

4.1.6   Fuses for the protection of circuits containing semiconductor devices

The energy let-through to a protected circuit during short-circuit conditions is related to the integral with respect to time of the square of the instantaneous current, i.e. $\int i^2\,dt$. For any circuit and piece of equipment there is a maximum value of this quantity which must not be exceeded if healthy equipment is not to be damaged. This quantity is designated I^2t and the withstand I^2t values of circuits have to be known or determined before suitable fuses can be selected. Further information on this topic is provided later, in Section 7.8.

The I^2t withstand values of semiconductor devices of given ratings are considerably lower than those of other components and circuits of corresponding ratings. Fuselinks used in circuits containing semiconductor devices must therefore be capable of operating more rapidly at given currents than fuselinks used in other applications.

Cartridge fuselinks of the same basic constructions as those described in Section 4.1 were developed and have been available for some time to protect semiconductor devices. These fuselinks, which are usually referred to as semiconductor

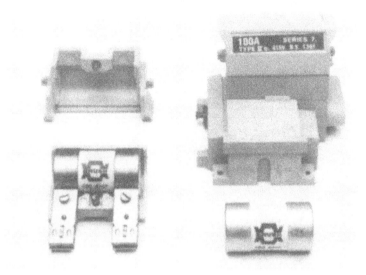

Figure 4.18   House service fuse

fuselinks, invariably have notched strip elements. The restricted sections have relatively small cross-sectional areas so that they operate at quite high temperatures at rated current, temperatures up to 250°C being quite common. Such a mode of operation limits the extra energy input needed to raise element temperatures to the melting point when short circuits occur on protected circuits and the durations of pre-arcing periods and the input I^2t values are reduced.

Continuous operation of elements at high temperatures would cause rapid oxidation of materials such as copper. This would clearly be unacceptable and therefore silver elements are invariably used in semiconductor fuselinks.

To maintain stable high-temperature operation of the restricted sections of elements, heat energy must be conducted away from them. Two measures are taken to achieve this objective. First, the restrictions are made relatively short so that the energy to be conducted away from each of them is limited. Second, the widths of the elements, i.e. the unrestricted sections, are made relatively large to enable them to run at low temperatures and thus to conduct sufficient heat energy from the restrictions.

As explained earlier, current interruption is effected in a fuselink when the total voltage across the arcs is sufficiently high relative to the system voltage. In practice the necessary total voltages are obtained by providing the required numbers of restricted sections.

The differences outlined above in the geometries of the elements provided in fuselinks used to protect semiconductor devices and those used in normal general-purpose fuselinks are illustrated in Figure 4.19.

To assist further in obtaining the performances required of semiconductor fuselinks, techniques are employed to ensure that their quartz filling material is very

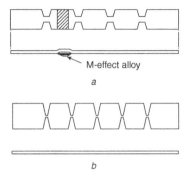

M-effect alloy

a

b

Figure 4.19   Fuselink elements
  a   Industrial fuselink
  b   fuselink (for semiconductor protection)

highly compacted. This constricts the arcs and reduces the I^2t let-through when short circuits are being cleared. This practice also increases the levels of direct current which can be interrupted by these fuselinks, an improvement which is important when semiconductor convertor equipment is to be protected.

To further reduce the I^2t values required by fuselinks of given current ratings when clearing short circuits, the grains of the quartz filling materials are often coated with an inorganic binder. This increases the contact areas between the grains and thereby increases the thermal conductivity of the filling material. This does not affect fuselink behaviour when short circuits occur, but increases the permissible rated currents.

Because the restrictions in the elements of semiconductor fuselinks operate at quite high temperatures when rated currents are flowing, it is essential that their bodies and terminations be able to dissipate considerable amounts of heat. They are, therefore, not normally mounted in enclosed fuse holders and it is recommended that they be placed in positions where there is adequate natural ventilation. In some cases, the elements are placed in two or more bodies mounted in parallel to increase the surface areas from which heat can be dissipated. All designs have very substantial terminations to enable heat to be conducted away from the elements.

Typical single- and double-body fuselinks are shown in Figure 4.20.

In some applications it is desirable to have a positive indication when a particular fuselink has operated. Special trip-indicator devices, which are electrically and usually mechanically coupled in parallel with the fuselink, are produced for this purpose. Such an arrangement is shown in Figure 4.21. Operation is initiated when the element in the fuselink melts. The current then transfers to the fine element in the trip-indicator device and this element melts quickly to release the plunger which is driven out by a spring. The plunger gives local indication of operation and, if desired, it may be used to operate a microswitch, which can then in turn operate a remote-warning device or initiate additional protective equipment.

Figure 4.20   Typical British semiconductor fuselinks

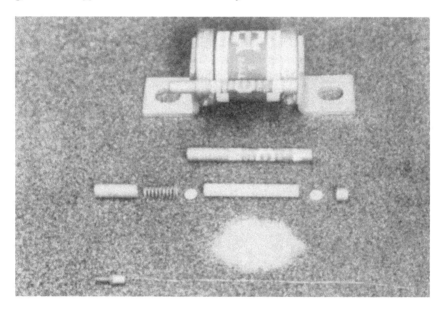

Figure 4.21   Trip indicator for semiconductor fuselink

In the UK, the majority of power semiconductors are used in three-phase and single-phase equipments operating at 240 V RMS per phase and fuselinks, dimensionally standardised to comply with *IEC 60269-4-1 Section IA*, are available for single-phase applications with ratings up to 900 A at 240 V. They are also available for use in three-phase circuits operating up to 660 V (line) in current ratings up to 710 A. The standard dimensions are as small as practicable because many fuselinks are used in applications where space is at a premium.

Non-standard fuselinks have been produced for use in very high-power three-phase equipment with current and voltage ratings up to about 4000 A and 7·2 kV (line).

Figure 4.22   Types of fuse-holders

A recent development has been the deposition of metal fuse elements onto a high-thermal-conductivity ceramic substrate. This permits good thermal conduction from the fuse element to the substrate which gives a higher continuous-current rating and enhanced withstand to normal repetitive overloads.

4.1.7   Other types

Many other fuselinks are in service, which include: fuselinks generally to *IEC 60269-2-1 Section II* but with special terminations, fuselinks for use in aircraft and special DC fuselinks for traction applications. Fuse-holders with non-standard cable terminations as shown in Figure 4.22 are produced and in addition unshrouded fuse-holders and open-type fuse bases are available for use in cubicles which are isolated when opened.

4.2   Semi-enclosed fuses

These are used in three-phase industrial applications where the system voltage does not exceed 240 V to earth and they are available in current ratings up to 100 A. They consist of a fuse-holder made up of a fuse base and carrier, the latter containing the element which is invariably in wire form. The elements are directly replaceable and of low breaking capacity, being typically capable of interrupting currents up to only 4000 A. Their breaking capacity is comparable with that of many miniature circuit breakers.

They are produced either with a porcelain base and carrier or with a moulded plastic carrier containing a ceramic insert through which the wire element passes. A typical industrial semi-enclosed fuse is shown in Figure 4.2.

For domestic applications, semi-enclosed fuses, cartridge fuses or miniature circuit breakers may be used in consumer units to provide overcurrent protection, and most modern units can be adapted to take any of them. An example of a domestic semi-enclosed fuse is shown in Figure 4.23.

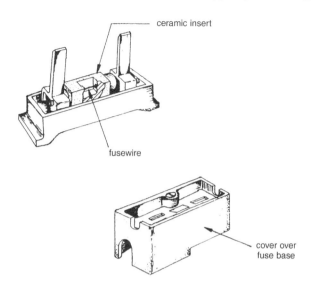

ceramic insert

fusewire

cover over
fuse base

Figure 4.23   Domestic semi-enclosed fuse

4.3   Continental European fuses

Three main systems are used on the continent of Europe and in many other countries throughout the world where continental European low-voltage equipment is generally used. These three systems are described in terms of the contact arrangements of the fuselinks. These are:

(*a*)   blade contact – 'NH' type
(*b*)   end contact or screw type – 'D' type
(*c*)   cylindrical cap contact.

Each type employs cartridge fuselinks and no equivalents of the British semi-enclosed fuses are produced.

4.3.1   Blade-contact-type fuses

These are generally referred to as NH fuses, NH being an abbreviation of Niederspannungs Hochleitungs which is German for low-voltage high-breaking-capacity. They are for use by authorised persons, mainly for industrial applications, and are used in factory distribution systems and also in the distribution cabinets of the ESI in power-distribution networks.

A range of these fuselinks is shown in Figure 4.24. The fuse elements are generally made from copper strip. The body is usually made of ceramic but high-temperature thermosetting plastic materials have been used. Bodies often have a rectangular outside cross-section with a circular longitudinal hole through them, and end plates, complete with the blade contacts, are attached to the body with screws. To allow the

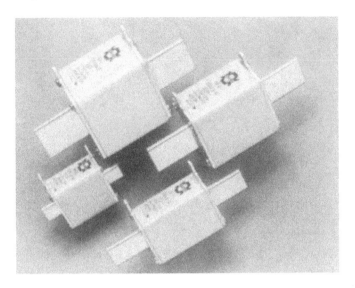

Figure 4.24   NH fuselinks

fuselinks to be mounted in close proximity to each other, even in the absence of insulating separators, the end plates are normally confined within the outside dimensions of the fuselink body. The blade contact surfaces are usually silver plated to assist in obtaining low-resistance connections even when the forces applied on the blades by the spring contacts into which they fit are relatively low.

Some fuselinks are produced with cylindrical bodies and these are allowed in the standard specifications provided that they meet the dimensional requirements.

NH fuselinks are generally available for applications up to and including 1250 A, for AC circuits operating at levels up to 500 V and DC circuits of voltages up to 440 V. Designs with restricted current ratings are available for 690 V AC systems. They usually incorporate operation indicators, a feature which is not now normally provided on low-voltage fuselinks used in the UK. The operation-sensing device is a fine wire which is connected in parallel with the fuse element. This wire normally carries only a tiny fraction of the total current passing through its fuselink but, when the element ruptures during clearance, the wire carries a much larger current, which causes it to melt and break very quickly. The wire is used to hold in a flag or plunger in one of the end plates. When the wire breaks, the flag or plunger is pushed out by the action of a spring and in this way an indication of operation is given.

In some installations, only the front of the fuselink is visible and there is therefore a requirement for an indicator at the centre of the front of the fuse body. In order to reduce the stocking of the two types of indicating fuselinks, designs are readily available which have a combined front and end indicator. An example is shown in Figure 4.25.

Unlike British designs, NH fuselinks are inserted into their fuse bases by a detachable handle which is made of plastic, a particular example being shown in Figure 4.26.

Figure 4.25   NH fuselink with combined front and end indicator

Alternatively, in a widely used simple design of fuse-switch, the cover of the switch acts as the fuse handle, the fuselinks replacing the normal switch blades and being withdrawn when the cover is opened or removed. This arrangement is shown in Figure 4.27.

Traditionally, the fuse bases have been much simpler than the British designs described earlier and were not shrouded to prevent accidental contact with live parts. However, in line with other low-voltage equipment, associated safety aspects have been addressed and fuse bases with covers and shrouds over the contacts are readily available. In addition, fuselinks with isolated gripping lugs are offered.

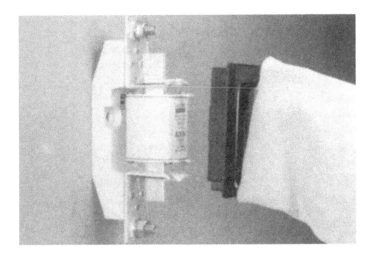

Figure 4.26   NH feeder pillar showing use of handle

Figure 4.27   NH fuse switch

For electricity supply industry applications, NH fuselinks are used in fuse distribution cabinets, including packaged substations and cable distribution cabinets. This
has led to the introduction of integral three-phase shrouded units which are mounted
directly onto the distribution bus bars. This system has become standardised in *IEC
60269-2-1* and is referred to as 'fuse rails'. A shrouded three-phase unit is shown in
Figure 4.28.

In view of the number of fuselinks in a distribution cabinet, it is important to
minimise the temperature rise within the cabinet and low power loss fuselinks are

Figure 4.28   Shrouded three-phase NH 'fuse rail' unit

desirable. In addition, if low power loss fuselinks are installed by a utility, then if the projected reduction in power loss is multiplied by the number of fuses installed then this can provide a measurable energy saving. As stated earlier, the 'standard' voltage rating for NH fuselinks is 500 V AC. However, most distribution systems outside of North America are based on a 400 V three-phase supply and consequently 400 V fuselinks with a lower power loss than 500 V traditional designs are available in the standard NH dimensions.

In Germany and associated markets, three high-current rating fuselinks are often placed as close as is practical to the secondary terminals of the distribution transformer. This gives added protection in the case of an accidental fault between these fuses and those downstream protecting the multi-way feeders. These fuselinks are given the designation gTr in Germany. However, to date they have not been introduced into the IEC standard. The 'nameplate' rating of these fuselinks and characteristics matches the kVA rating of the transformer, thus simplifying the selection of the appropriate fuselink. The gTr fuselinks allow to run the transformer at 130 per cent of its rated current over a period of 10 h, which is sufficiently long to cover the daily high-load period of power utilities.

4.3.2   End-contact or screw-type fuses

As indicated in Chapter 1, this is a very old fuse system, often referred to as the Diazed type or in some countries by the abbreviated form of this: 'Zed' type. Diazed was derived from 'diametral abgestruft', that is 'diametral steps'. The term 'bottle' type is also used, this name clearly stemming from the characteristic shape of the fuselinks, an example of which is shown in Figure 4.29. The official designation, that should be used, however, is 'D' type which is derived from Diazed.

Production of these fuses is perpetuated by the continuous need to supply replacements for some of the very large number which have been installed over many years. A sectional view of a typical fuselink is shown in Figure 4.30. They are mainly produced with ratings up to 63 A for use in AC circuits operating at levels up to 500 V. The limitation in current rating is largely caused by the difficulty of producing satisfactory contact with the holders, rather than with shortcomings in the fuselink design. Higher ratings, up to 200 A, have been produced but they have not proved popular.

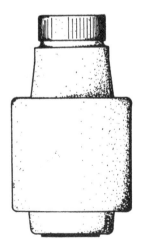

Figure 4.29   'D' type fuselink

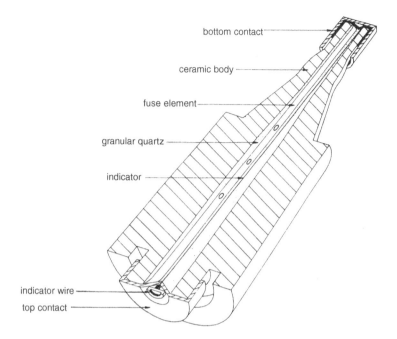

Figure 4.30   Sectional view of 'D' type fuselink

The fuselinks contain strip elements of copper or silver-plated copper and they are filled with granular quartz. The bodies, which are made of ceramic material, often have much thicker walls than British fuselinks of equivalent ratings which tends to assist with heat dissipation because porcelain has a higher thermal conductivity than granulated quartz. Each fuselink is fitted at the ends with cylindrical contacts made of brass, usually nickel-plated, and they are often of two different diameters. They usually have grooves in the ends to ensure good contact when they are fitted in the carrier.

The fuselinks are fitted with operation indicators which generally take the form of a button head that is pushed out through the end contact by a weak spring when the fine-wire operation-sensing device in parallel with the main element melts and no longer provides restraint. The button head is visible through a glass window in the screw cap.

A standard holder of the form shown in Figure 4.31 is produced to accommodate a range of 'D' type fuselinks. A range of gauge rings, with various internal diameters and coloured ends to indicate the maximum ratings of fuselinks which will fit into them, is available. The appropriate ring is placed into a fuse-holder to ensure that a fuselink of too great a rating for the circuit being protected may not be installed. The fuselink is inserted before the screw cap is screwed to the fuse-holder and this produces forces between the fuselink end contacts and the spring contacts in the screw cap and fuse socket.

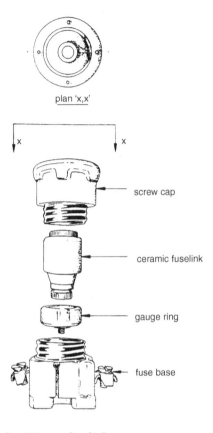

plan 'x,x'

screw cap

ceramic fuselink

gauge ring

fuse base

Figure 4.31   Holder for 'D' type fuselink

A dimensionally smaller range of fuses, designated DO, has been produced by at least two continental manufacturers. These are used in some countries, but in others, only the standard 'D' type is allowed in the interests of standardisation.

DO fuse-switch-disconnectors have recently been introduced which offer the following features:

- It can only be switched in when the DO fuselink is firmly screwed in position, giving high contact pressure.
- In the off position and the fuselink *in situ*, it isolates the fuse from the supply.
- In the off position and the fuselink is unscrewed, the fuselink is isolated from both the supply and the load. No dangerous reverse voltages are therefore possible.
- Independent manual operation of the switch.
- Ganged three-phase units with 'DIN rail' mounting.

An example is included in Figure 4.32.

Figure 4.32   DO products including fuse-switch-disconnectors

4.3.3   Cylindrical-cap-contact fuses

Fuse systems incorporating fuselinks with cylindrical bodies and ferrule end caps are widely used in France and associated countries for domestic and industrial applications. The fuselinks are filled with quartz and usually have ceramic bodies. They are available with operational indicators if required.

For domestic applications, the fuselinks are produced in a range of diameters and lengths each having its own unique dimensions to prevent incorrect replacement after operation.

In *IEC 60269-3-1* these fuselinks are referred to as Type A to differentiate these from Type B, used in the UK and associated countries, and an old Type C, which was historically used in Italy. The Type A fuselinks are standardised in the following ratings:

250 V AC   up to 16 A

400 V AC   up to 63 A

For industrial applications, cylindrical fuselinks are available in the following standardised ratings:

400 V AC   up to 125 A

500 V AC   up to 100 A

690 V AC   up to 50 A

A fuse-holder widely used internationally for these fuselinks is shown in Figure 4.33. It has an integral hinged fuse carrier, which allows the fuselink to be inserted in a safe manner. It can be mounted on a standard 'DIN rail' in multiple linked modules such as three-pole and neutral.

Fuse combination units incorporating cylindrical fuselinks are available for industrial use. Strikers, operating in a similar way to operation indicators, may be incorporated in these fuselinks. When a fuselink melts, the striker moves out through

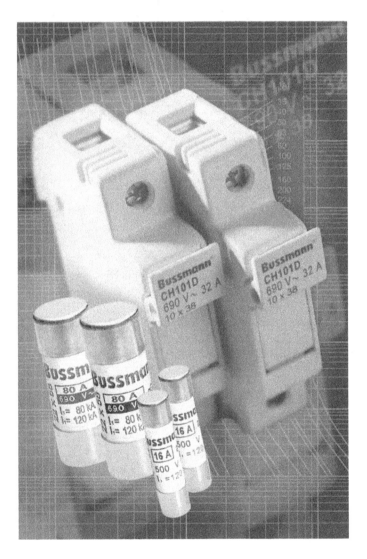

Figure 4.33   Cylindrical fuse-holder and fuselinks

the end cap and actuates a microswitch which may initiate an alarm. In three-phase designs the strikers can operate on a common trip rod which then actuates a microswitch which can energise the trip coil of the switch, or alternatively it can de-energise the holding coil to effect three-phase clearance. This feature is valuable in certain applications, such as the protection of motors, where single-phase operation is unacceptable.

4.3.4   Semiconductor fuses

The majority of these fuses are used to protect power semiconductors and usually rated at 690, 1000 or 1250 V AC, but other voltage ratings will be found. Square-ceramic-body designs are very popular, the body lengths being shorter than those of fuselinks used for industrial applications. End terminations suitable for bolted connections with fixing centres at 80 or 110 mm are widely used, but alternative versions with tapped holes in the ends are available. The latter design has the advantage of being more compact and is often used for ratings above 1000 A.

The overall dimensions of semiconductor fuses are specified in *IEC 60269-4-1, Section 1B (DIN 43653)* and a selection of the designs available are shown in Figure 4.34.

Many semiconductor fuses used in Continental Europe incorporate operation indicators. These are similar to those used in industrial fuselinks and they may be either at one of the ends or in the central regions of the bodies. As well as giving local indications, these devices may be adapted to operate microswitches so that remote indication of fuse operation may be provided. Examples of these fuselinks are shown in Figure 4.35.

4.4   North American fuses

The internal demand for electrical equipment is so great in the USA that special fuse systems have been developed for the home market. These systems are also used to some extent in Canada, where British practice is also popular, and in Central America, certain parts of South America and the Philippines. Outside of these territories these fuses are found in equipment exported from the USA.

4.4.1   Industrial fuses

Several types of fuses are produced for industrial applications with alternating current ratings up to 6000 A and the ability to operate under test in circuits with maximum voltages of 250 or 600 V.

The various fuses are divided into classes, the main classes being:

Class R – up to 600 A, 250 and 600 V
Class H – up to 600 A, 250 and 600 V
Class J – up to 600 A, 600 V
Class L – above 600 A and up to 6000 A, 600 V.

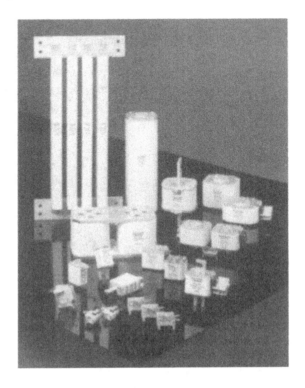

Figure 4.34   Continental European semiconductor fuselink

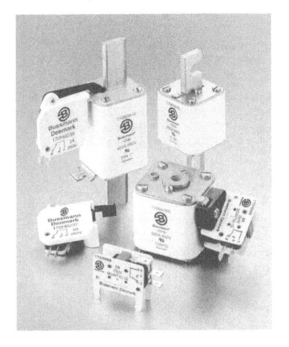

Figure 4.35   Indication of operation

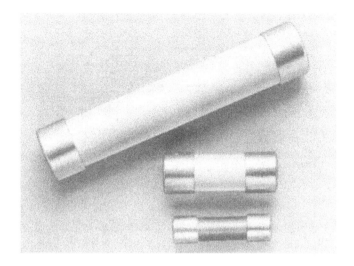

Figure 4.36   North American cylindrical fuses

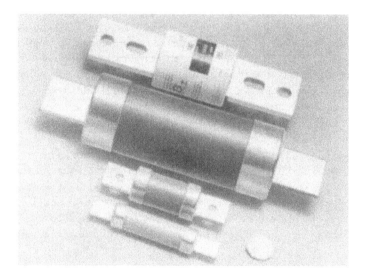

Figure 4.37   North American tag-type fuses

Fuselinks with current ratings up to 60 A are generally fitted with cylindrical end contacts and they have the external appearance shown in Figure 4.36. Fuselinks for higher ratings are provided with either blade-type terminations for mounting in spring contact or blade terminations containing holes or slots which allow bolted connections to be made, examples being shown in Figure 4.37.

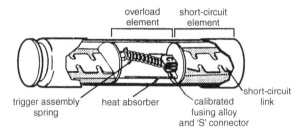

Figure 4.38   Class R dual-element fuselink

The class R fuselinks have a high breaking capacity of a minimum 200 kA. They are cartridge fuselinks with cylindrical bodies which are usually made of materials ranging from vulcanised fibre to glass cloth impregnated with resin, or pultruded thermoset polyester, these materials being used for industrial fuse bodies in the USA. A number of different time/current characteristics are available which are governed by the element design. The most popular industrial fuselink in the USA is the class R dual-element time delay. This type of fuselink operates slowly at currents in a range above the minimum fusing level but nevertheless clears rapidly when carrying very large currents. This characteristic is appropriate for protecting equipment such as motors where there is a transient surge of current on starting. A method of obtaining this characteristic is to split the fuse element into three sections. The two outer sections, which are of the conventional strip form with restrictions, are surrounded by an arc-quenching material such as quartz or, in one design, calcium sulphate. It is these sections which operate when high short-circuit currents flow. The elements are generally made from copper but silver is sometimes used when the let-through energies ($\int i^2 \, dt$) under high faults are to be kept to a low level. The centre section is unusual in that it often has two metal portions of large thermal mass connected together with low-melting-point solder. At currents somewhat above the minimum fusing level the solder melts, its temperature rise being slowed by the adjacent masses of metal. To provide an adequate break, the two metal parts are pulled away from each other by springs when the solder melts. The centre section does not contain filling material because this would prevent movement of the metal parts and, of course, it is not required for arc-quenching purposes at the current levels which are cleared by the centre section of the fuselink. The whole mechanism, which is often referred to as a 'dual-element' arrangement, is based on the same principle as the fuses produced in 1883 by Boys and Cunyngham, to which reference was made in Chapter 1. An example of fuselinks of this type is shown in Figure 4.38. Other types of class R fuselinks use the M-effect principle.

The class H fuselinks, sometimes referred to as 'code' fuses, have a low breaking capacity of 10 kA. They employ cylindrical fuse bodies of the same materials as the high-breaking-capacity fuselinks and the connection arrangements and dimensions also correspond so that they are interchangeable, unless rejection fuseclips are fitted.

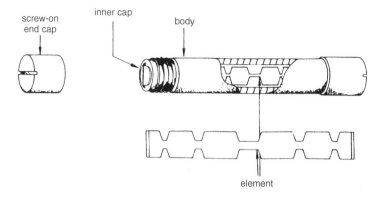

Figure 4.39   Renewable fuselinks

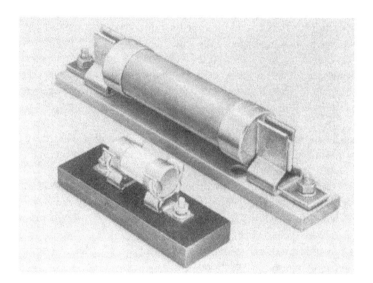

Figure 4.40   North American fuse bases

The first type, which is known as non-renewable or 'one time', incorporates copper elements and filler, whereas those of the second type, designated renewable, do not contain filling material and use replaceable zinc elements with restricted sections. The end caps and terminations of this latter type are removable to allow the elements to be replaced when necessary. Details of the general construction of renewable fuselinks can be seen in Figure 4.39.

Both classes R and H fuselinks are usually mounted in simple, unshrouded fuse bases; some examples are illustrated in Figure 4.40. In Canada the class R fuse is often referred to as a high-rupturing-capacity (HRC) class H fuse.

Figure 4.41   Rejection features in class R fuselinks

As stated earlier, class R and class H fuselinks have the same dimensions and are, without rejections clips, physically interchangeable. As a result, care must be exercised in service to ensure that class R fuselinks are never replaced by those of class H, because of the low breaking capacity of the latter. To eliminate the possibility of errors in the future, class R fuseblocks have a rejection feature (see Figure 4.41) that will prevent the insertion of class H fuselinks. This will help to ensure the replacement of 200 kA or greater fuselinks.

The class J fuselinks are more compact than classes R and H. They have a breaking capacity of a minimum 200 kA and have a high speed of operation under high fault levels to give low let-through values (I^2t).

The class L fuselinks are used for all applications for current ratings above 600 A. They have a minimum breaking capacity of 200 kA.

Class T has smaller dimensions than the class J and voltage ratings of 300 and 600 V.

Further information on the above and other classes of fuses can be found in the *UL248* series of standards, see Section 8.4.1.

The class J and class L fuselinks have recently been introduced into the IEC 60269 series of standards and can be found in *IEC 60269-2-1 Section VI*.

Since traditionally the North American fuses are of the 'open type', there is a need to give authorised persons increased safety to avoid touching live parts, hence there is an increasing trend in North America to provide covers or shields over the live parts; see Figure 4.42

This has also led manufacturers to produce new fuse designs for the popular ratings. An example is a fully shrouded fuseholder for up to 60 A in the class J size. More radical developments have been made by manufacturers including a new fuse system, the cube fuse. One of the main features of the cube fuse is that it eliminates the need for a fuse carrier or replaceable handle. The fuselink is fitted or withdrawn with finger pressure into or out of the fuse base. The 'cube fuse' is shown in Figure 4.43.

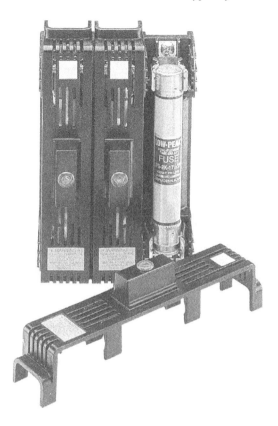

Figure 4.42   Shielded North American fuse

4.4.2   Domestic fuses

For domestic, commercial and light-industrial distribution applications, class G fuses are available for a rated voltage of 480 V and alternating-current ratings up to 60 A with a breaking capacity of 100 kA.

In North America, unlike the UK, fuses are incorporated in the sockets from which the supplies are obtained for the individual items of equipment used in domestic premises. Because the fuselinks are screwed or plugged into their holders they are known as 'plug' fuses. They generally have rated currents up to 30 A and are for use in AC circuits operating at 125 V to ground.

Three types are produced, these being:

(*a*)   the ordinary plug fuse (Edison-base)
(*b*)   the dual-element plug fuse (Edison-base)
(*c*)   the type S plug fuse (Fustat-base).

The ordinary plug fuse which is shown in Figure 4.44 contains an element in either strip or wire form. The element can be seen through a window set in the end of the body.

Figure 4.43   'Cube fuse'

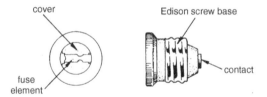

Figure 4.44   Plug fuse

The body, which does not contain filling material, is usually made of glass, porcelain or high-temperature plastics. It has a metal contact in one end and a threaded brass section around it which acts both as the second contact and the means for screwing the fuselink into its base or holder. The dual-element plug fuse has the same basic construction as the ordinary plug fuse except that it contains two copper strips which each have one of their ends soldered to the other. A spring is incorporated to separate the strips when the solder has melted as a result of the passage of overloads. The principle is similar to that on which the class R fuselink, described above, is based. The effect is the same, namely that operation is not produced by harmless transitory surge currents but rapid clearance is nevertheless obtained in the event of very high currents flowing during short circuits.

Both of the above types of plug fuse are fitted into standard Edison-screw bases. It will be realised that it is possible for users to make contact with live parts when the fuselink is removed and a fuselink of any rating may be fitted into the base.

The type S plug fuse is similar to the dual-element fuse but it is designed for use with an adaptor which performs a similar role to the gauge rings used with the 'D' type fuses, described earlier in Section 4.3.2. The adaptor has an external screw thread which will mate with the internal thread of a standard Edison-screw base, and a tang with several different internal threadings, and in this way the easy insertion of an over-sized fuselink is prevented.

4.4.3 Semiconductor fuses

The fuselinks provided in North American circuits containing power-semiconductor devices are similar to those used in industrial applications. As in the UK, the special performance requirements have been obtained by using suitable materials and element geometries.

The dimensions of these fuselinks have not been standardised by any national body but industry-standard envelopes have evolved by usage in the AC RMS voltage ratings of 130, 250, 500 and 700. In order to minimise the future proliferation of dimensional variants, popular examples of dimensional system are given in *IEC 60269-4-1*:

Section IC for bolted connections.
Section IIA for cylindrical contact caps.
Section IIB for flush end connections.

Figure 4.45   North American semiconductor fuses

Ferrule type fuselinks are normally used in circuits with ratings up to 30 A. Fuselinks with terminations suitable for making bolted connections are also produced and these are used exclusively in circuits with high current ratings. These latter fuselinks are generally smaller than industrial fuselinks of corresponding ratings, and they are not therefore interchangeable. A selection of the North American fuses is shown in Figure 4.45.

4.5   Fuses for telecommunication power systems

The nominal voltage used in the United States for telecom wireless systems is 24 V DC and for wireline systems is 48 V DC with excursions to 57 V DC. The wireline voltage in some European countries can be as high as 75 V DC. Power is supplied by either a rectifier or stand-by batteries which are normally of the lead-acid variety. The rectifiers are of the current limiting type but not so with the batteries. Large central offices can have a full load requirement of over 5000 A and often use many parallel

Figure 4.46   Fuses, fuselinks and disconnect switches specifically for telecom applications

strings of batteries. Short-circuit current from such systems can reach very high levels approaching 100 kA in some cases.

The voltage levels employed in the telecom industry are not sufficiently high so as to cause death from electric shock, however based on the potentially high level of short circuit current available, energy hazards, fire and mechanical and heat hazards can easily cause severe injury to personnel and/or damage to equipment. Compounding this problem is the industry practice of working on these systems while they are still up and running.

Fuses are used in these power systems to limit any damage due to short-circuit current. Since modern fuses designed for the telecom industry are of the current limiting type, the cables are better protected, arcing within the system is minimised and consequently, the potential for fires to occur is also minimised. Additionally, selective co-ordination of the various fuses throughout the system is more easily achieved which will reduce system downtime.

Fuses have been used for many years for telecom power system protection. Prior to the late 1980s the industry used standard industrial type AC fuses on their DC circuits, but since that time, new fuses and associated disconnect switches have been designed specifically for the industry. Initially the focus was on the industry in the USA and consequently early fuse designs have a voltage rating of 60 V DC and an interrupting rating of 100 kA. Today many different styles exist which now go up to 80 V DC, high enough to cover European requirements. These new fuses are physically much smaller than the older types and when used in conjunction with their associated disconnect switch in applications where multiple units are to be used, they offer exceptional space savings.

All the new fuse products possess full range protection capabilities from low overloads for cable protection to high short-circuit currents. No listing agency, such as Underwriters Laboratories (UL), has an industry standard for such fuses, so consequently in the USA they are all UL component recognised.

A selection of special fuses, fuselinks and disconnect switches are shown in Figure 4.46.

Chapter 5

Constructions and types of high-voltage fuses

By definition high-voltage (HV) fuses are for use in AC systems operating at frequencies of 50 and 60 Hz with rated voltages exceeding 1000 V.

These fuses fall into non-current-limiting and current-limiting classes, the latter being used exclusively in some countries, although both are used in other countries, including the UK.

Each fuselink should ideally be able to interrupt safely all currents from that needed to melt its element or elements up to its maximum rated breaking capacity. Those current-limiting fuselinks which can meet this condition are categorised as 'full range'. As explained in Section 5.2, many current-limiting fuselinks produced for use in HV circuits are not, however, able to provide the above performance, but they are nevertheless suitable for a wide range of applications. Such fuselinks may be categorised as 'partial range' or back-up fuses.

Following the pattern of the previous chapter, descriptions of the constructions of HV fuses produced and used in the UK are given first and then the practices in other countries are stated, particular attention being given to significant differences in design.

5.1 Non-current-limiting fuselinks

These fuselinks have short elements and include means for lengthening the short arcs initially set up when the pre-arcing period ends. Even so, the arc voltage is much lower than the system voltage and for this reason the current limitation produced tends to be insignificant. They provide 'full range' performance and operate, as do all current-interruption devices, by preventing arc re-ignition after a current zero.

Because the breaking performance is not greatly affected by the element parameters, the latter may be varied to provide different time/current characteristics. As an example, elements of different lengths could be used; a very short element, because of the high conduction of heat energy to its ends, has a small cross-sectional area for a given long operating time, at a particular current, but it would, as a result, operate very quickly at high currents, heat conduction in the short time available then being low. In practice, elements are made of tin wire in many cases, but tinned-copper wire is often used when fast operation is required.

The elements are subject to spring tension and therefore many designs incorporate a fuse wire in parallel with a strain wire of high tensile strength and high resistivity.

Details of the two main types of non-current-limiting fuses in use in the UK are given in the following sections.

5.1.1   Expulsion fuses

These fuselinks contain a short element of tin or tinned-copper wire in series with a flexible braid. These items are mounted in a fuse carrier incorporating a tube of organic material, usually closed at the top with a frangible diaphragm, and containing a liner of gas-evolving material such as fibre. The fuse element carries a closely fitting sleeve of gas-generating fabric. The flexible braid is brought out of the open lower end of the tube and held in tension, by a spring attached to the lower end of the fuse base.

A cross-section through a typical fuselink is shown in Figure 5.1 and three popular physical connection arrangements, designated button-head, double-tail and universal, are illustrated in Figure 5.2. It will be seen that these all have a braid at the lower end, it being the upper terminations which differ.

The fuse carrier has pins at the lower end which act as a hinge when it is mounted in the lower contact of the fuse mount. In the service position the fuse carrier is tilted from the vertical, as can be seen from Figure 5.3, which shows a complete assembly.

When the fuse element melts during operation, the release of the spring tension disengages a latch that allows the fuse carrier to swing down by gravity. This provides

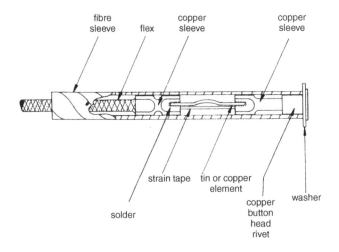

Figure 5.1   Sectional view of an expulsion fuselink

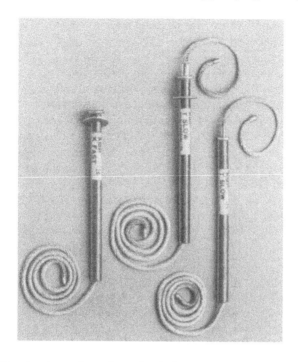

Figure 5.2   Terminations of expulsion fuselinks

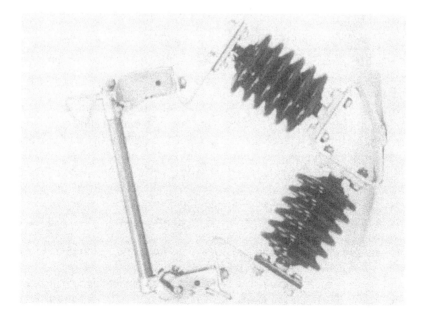

Figure 5.3   Expulsion fuse assembly

isolation and prevents discharges along the tube which could lead to tracking. It also indicates that operation has occurred.

These fuses are able to break a wide range of fault currents. When interrupting a small fault, the arc is extinguished within the fibre sleeve around the element, but at high currents this sleeve bursts and the arc is extinguished within the liner of gas-evolving material. The diaphragm at the upper end of the fuse carrier ruptures when the current being interrupted is sufficiently high and thus double-end venting is provided, so relieving the fuse-carrier tube of excessive pressure.

These fuses are available for use in three-phase circuits with current and line voltage ratings up to 100 A and 72 kV, respectively. Their maximum breaking capacity is typically limited to 150 MV A.

Expulsion fuses are for outdoor use only and they may be replaced using a pole from the ground. The pole can also be used to swing down the carrier of a healthy fuse for the purpose of isolation. It is necessary to check that the circuit is not carrying current when this latter operation is to be performed, unless the equipment is arranged so that load current may be interrupted by the contacts. In some designs this is done by including circuit-breaker-type arc chutes around the contacts which separate. An alternative, which is sometimes adopted, is to use fuselinks which are so designed that the element may be snapped mechanically by an operator inserting a pole into it from the ground. The fuselink then functions in the normal way to clear the circuit, after which the carrier may be swung down safely.

5.1.2   Liquid fuses

In the earliest non-current-limiting fuses, the arcs were quenched in a liquid and this principle has been used for many years to produce fuselinks. Many of them are in use in the UK. They have a body consisting of a glass tube with metal ferrules at each end and within it is the element. This is normally of silver strip or wire, with a strain wire across it except for low current ratings, say 10 A and below, in which a wire is used as the element and this is made strong enough to make a separate strain wire unnecessary. In all designs the element is positioned near the top of the tube so that it is shielded from corona discharge by the upper ferrule. The element is held in tension by a spring anchored to the lower ferrule and the tube is filled with an arc-extinguishing liquid, usually a hydrocarbon. When the element melts during operation, the spring collapses and the arc is extinguished in the liquid. To relieve the tube of excessive pressure, a diaphragm on the upper ferrule is ejected, except under very moderate conditions. Provision is usually made to allow the user to recharge by inserting a new fuse-element assembly and new liquid. Figure 5.4 shows a cross-sectional view of a typical liquid fuselink. These fuses are only in use outdoors and provision is made for removing from and replacing the fuselinks into their mountings by pole operation from the ground, a bayonet-fixing arrangement being usual. They are mostly used for the protection of 11 and 33 kV pole- or pad-mounted transformers on rural systems and also for spurs feeding a number of transformers.

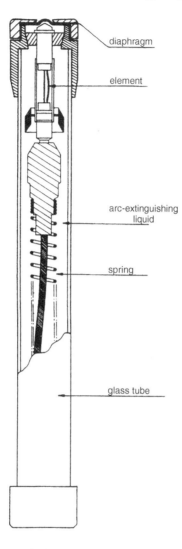

diaphragm

element

arc-extinguishing
liquid

spring

glass tube

Figure 5.4   Sectional view of a liquid fuse

The breaking capacity of these fuses is limited to values below those of expulsion fuses and although they have given, and indeed still are giving good service, they are no longer recommended for new installations.

5.2   Current-limiting fuselinks

As stated earlier, many of the current-limiting fuselinks which are produced are in the back-up or partial-range category. Their constructions and behaviour are described in the following sections and then fuselinks in the full-range category are described in Section 5.2.3.

5.2.1   Constructions of back-up or partial-range fuselinks

These are of the cartridge type, constructed basically as described in Section 4.1. The special features associated with HV fuselinks are outlined below.

Fuselinks with low rated currents have silver-wire elements. Those with higher ratings have elements made from silver strip, with restricted sections at regular intervals along their lengths. These are produced by punching out appropriately shaped notches or holes. All elements have low-melting-point metals applied to them to enable the M-effect to be obtained.

It was explained in Chapter 3 that the speed of arc extinction depends on the voltage dropped across the arcs. It was also explained that the sum of the voltages across the arcs tends to limit the current rise initially and thereafter to cause the current to reduce. For this process to be effective in HV circuits, the total voltage across the arcs in a fuselink must be high, and to achieve this requires either a very long arc or many short ones, the latter being somewhat more effective because of the anode- and cathode-fall voltages associated with each arc.

In consequence and because the total length of break, after operation, must be sufficient to withstand the system voltage and provide adequate isolation, the elements in high-voltage fuselinks are long and have many restrictions. These lengths exceed the lengths of the fuse bodies needed to prevent external flashovers between the end caps at voltages of about 3·6 kV and above and therefore to make compact designs at these voltage levels the elements are accommodated in a helical form by winding them on formers. These are usually made of a refractory material and have a star-shaped cross-section. Such a former is in contact with the elements wound on it only at a number of localised points, and maximum contact between the quartz filling material and the elements is thus achieved. The inclusion of formers assists in obtaining consistent performance because they ensure that elements are correctly spaced from the bodies and in multiple-element fuselinks the individual elements can be equally spaced from each other. The former, together with one or more parallel-connected elements, is mounted in the fuse body which is cylindrical and usually made of ceramic material which is sometimes reinforced by adding a glass-fibre covering. The complete arrangement of a typical fuselink is shown in Figure 5.5.

It is common practice in the UK for HV fuselinks to be immersed in oil in various pieces of equipment, such as ring-main fuse-switch units. Several advantages arise because of the cooling effect of the oil, which allows a given current rating to be obtained with a smaller cross-section of element than is necessary in a fuselink which is to be used in air. The smaller element operates more rapidly at very high current levels and the fuselink can be made physically smaller and cheaper.

Fuselinks which are to be immersed in oil must be sealed and tested to ensure that they will prevent the ingress of oil. They are normally fitted with cylindrical cap contacts of 63·5 mm diameter and are produced in two basic overall lengths of 254 and

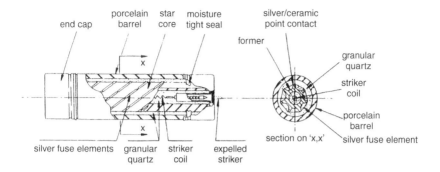

Figure 5.5   Construction of typical high-voltage fuselink

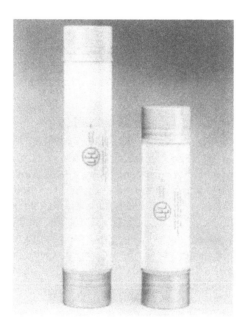

Figure 5.6   Fuselinks suitable for oil immersion

359 mm. The caps which are of copper or brass are tin or nickel plated. Examples of these fuselinks are shown in Figure 5.6.

These fuselinks, which have references F01 and F02, are available for use in three-phase circuits, principally at 11 or 13·8 kV, and in current ratings up to about 100 A. Voltage ratings from 3·6 up to 24 kV are also available. They are capable of breaking duties in excess of 250 MVA.

In recent years, over-zealous safety regulations and more restrictive maintenance budgets are resulting in lower usage of oil-immersed equipment. In such cases air type current-limiting HV fuses may be included in the T-off circuit of an SF$_6$ ring

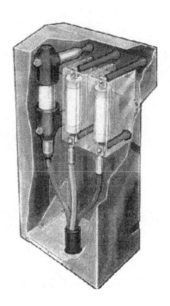

Figure 5.7   Fused end box

main unit. Alternatively full range fuses (see Section 5.2.3) may be fitted as sole protection, e.g. in a fused end box attached to the side of the transformer tank or in a simple non-striker tripped switch unit. An example of a fused end box is shown in Figure 5.7 which illustrates two types of insulation methods for the fuselinks.

High-voltage fuselinks, given the reference TA3, are produced for use in air and these generally have brass or copper spade terminations with holes in them to allow bolted connections to be made. These fuselinks are available for use in three-phase systems principally at 11 kV but ranging in voltage rating from 3·6 to 15·5 kV and with current ratings up to 100 A or more.

Fuselinks with cylindrical cap contacts of either 50 or 76 mm diameter and having references FA1–FA5 are produced in lengths of 359, 565 and 914 mm, these dimensions being necessary because of their high levels of rated voltage, up to 72·5 kV. They are for use in air. Typical examples of FA1–FA5 and TA3 fuselinks are shown in Figure 5.8.

Fuselinks used for the protection of three-phase motors generally have elements which are corrugated in the manner shown in Figure 5.9. This enables them to withstand the cyclic mechanical stresses which can arise during periods when the motors which they protect are being started and stopped repeatedly. These fuselinks generally contain a greater number of wider elements than type TA3 fuses, a combination which enables a high operating current at about 10 s to be obtained.

Designs for use at voltages up to 7·2 kV do not usually contain a former on which the elements are wound and supported. This increases the space available for elements in particular body sizes and thus makes it possible to include a greater number of

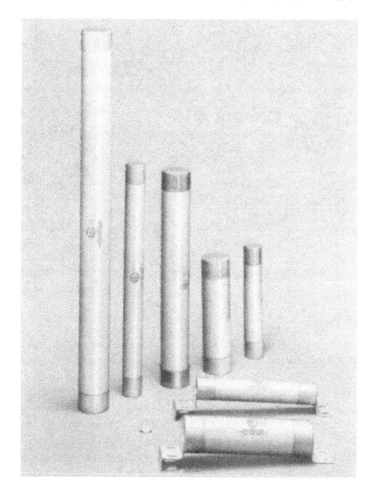

Figure 5.8   Fuselinks for use in air

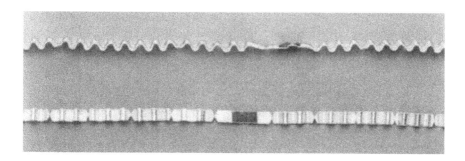

Figure 5.9   Corrugated elements

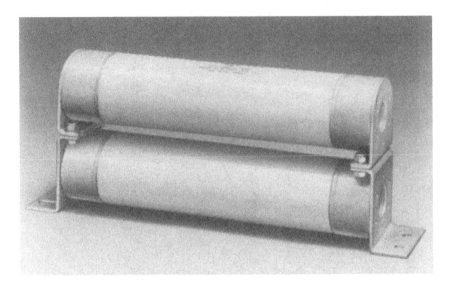

Figure 5.10   Group of two fuselinks for motor-circuit protection

elements. Motor-protection fuselinks are sometimes mounted in groups of two or three in parallel to achieve the high current ratings needed and bolted connections are usual. Typical voltage ratings are 3·6 to 7·2 kV with current ratings up to about 500 A. A group of two fuselinks is shown in Figure 5.10.

Many fuselinks for use in air outdoors are fitted with weatherproof seals.

Most of the above fuselinks can be provided with strikers to give indication of operation or, more important, to trip associated switchgear. As stated earlier, this latter feature is often essential in three-phase circuits as continued operation after a fuse has opened to clear a single-phase fault might be very harmful. The energy to drive the strikers may be derived from springs or small pyrotechnic devices, the latter being favoured by British manufacturers. In this type, which can be seen in Figure 5.5, a cylinder in the end of the fuselink houses the striker pin and a small charge, of gunpowder, into which an ignition wire is centrally placed. This wire and a series-connected high-resistance fuse wire are connected in parallel with the main fuse elements. After the main elements have melted, the voltage across the fuselink drives sufficient current through the ignition wire to heat it and ignite the gunpowder. The resulting explosion drives the pointed end of the striker through the fuselink end cap with sufficient force to operate fuse-switch trip mechanisms directly. The penetration of the striker through the end cap is clearly an indication of operation, and it may be used solely for this purpose.

The use of fuses to protect electromagnetic voltage transformers, which supply instruments and relays, is discussed in Section 7.6. The fuses are included in series with the high-voltage windings which normally carry very low currents by power-system standards. Whilst it would be desirable to use fuses with very low rated and minimum-fusing currents to give protection in the event of interturn faults in either set of windings, such fuselinks would have delicate elements that might break because of vibration or for other non-electrical reasons. This could de-energise relaying equipment and jeopardise major plant on the network. The risk cannot be taken and it is therefore accepted that the fuses role is for circuit isolation rather than transformer protection, the common rating available in the UK is 3·15 A. The construction of these fuselinks is similar to that of the conventional current-limiting fuselinks described earlier in this chapter. Because of their low-rated current they have only one element or two parallel-connected elements of silver wound on ceramic formers. They are produced for use in three-phase circuits with line voltages ranging from 3·3 to 33 kV and have cylindrical cap contacts or screwed end studs for connection purposes, examples being shown in Figure 5.11. They are not provided with strikers or operation indicators. Both sealed types for use in oil or outdoors in air, and unsealed designs for use indoors in air, are available.

The common practice of mounting low-voltage fuselinks in fuse carriers and bases is not adopted for high-voltage fuselinks because of their relatively large dimensions. The arrangements vary with the different applications and special mountings are often produced to enable complete fuse units to be incorporated within pieces of equipment.

Figure 5.11   Voltage-transformer fuselinks

Figure 5.12   Fuses mounted in ring-main unit

As an example, the fuselinks in some ring-main units are clipped into a withdrawable three-phase carrier which is mounted in the same oil-filled chamber as the switch mechanism. An illustration of this arrangement is shown in Figure 5.12.

5.2.2   Current-interrupting abilities and categories of partial-range fuselinks

When a high-voltage fuselink with one or more conventional parallel-connected elements is clearing a fault current, short breaks are formed at the element restrictions and short arcs result. The process continues until the burn backs have a total gap length which can withstand the recovery voltage present after the arcs have extinguished at a current zero. The total gap lengths needed in HV applications are clearly considerable, and the times required to melt and vaporise sufficient element material are correspondingly long. Consequently, only at currents above a certain level is the rate of lengthening of the arcs in any particular fuselink sufficient to so limit the arcing time that the temperature of the filling material is prevented from reaching a level at which its arc-extinguishing properties are so reduced that the fuselink would fail to clear. There is thus a range of currents above the minimum fusing level within which satisfactory clearance will not be effected. This situation, which does not arise with low-voltage fuselinks, can be accepted for some applications as stated later, but it is nevertheless undesirable and manufacturers sought to lower the minimum safe clearance currents. Experiments demonstrated that improvements were obtained by using large numbers of parallel-connected elements of small cross-sectional area rather than a smaller number of thicker elements. The improvement occurred because all the short arcs set up at each of the restrictions when a relatively low fault current is being cleared do not continue to burn in each of the parallel-connected elements. In practice they tend to commutate around the elements, the gaps in those elements, which are arcing, extending rapidly because of the high current densities in them. The voltages across these gaps rise to levels where the now shorter gaps in other elements ionise and cause the previously burning arcs to extinguish. This process shortens the time taken for current interruption to be effected. Performance was also improved by employing elements with long restricted sections of small cross-sectional area, which again increased the current densities.

It was nevertheless difficult to produce fuselinks which could safely clear all currents above their minimum fusing currents up to their rated breaking capacities, and therefore three internationally recognised categories of HV fuselinks were introduced, namely:

Back-up	Fuses in this category must be able to interrupt all currents between a minimum value specified by the manufacturer and the full rated breaking capacity.
General purpose	Fuses in this category must be able to interrupt currents from the rated breaking capacity down to the level at which the operating time is 1 h.
Full range	Fuses in this category must be able to interrupt all currents from the rated breaking current down to the smallest current which causes the fuse elements to melt.

It is the practice of the UK Electricity Supply Industry and of supply authorities in numerous overseas territories influenced by UK practice, e.g. Australia, South Africa, India and the Middle and Far East, to use fuse-switch ring-main units.

In the UK, such fuse-switch ring-main units of 250 MVA rating at 11 kV are covered by Electricity Supply Industry standard *ESI41-12*, which stipulates that the fuselinks must be provided with strikers which trip the switch instantaneously when one or more fuselinks operate.

This not only prevents single-phasing of any motors fed from the transformers but, equally important, eliminates the possibility of trouble if the equipment is subjected to a fault current less than the minimum breaking current of the fuse.

Typically, the total time to trip the switch from inception of arcing in the fuselink may be only 30–50 ms whereas it will generally take at least ten times as long as this for any persistent low-level arc within the fuse to cause any trouble.

It is, of course, only for fault currents below the stated minimum-breaking current for the fuse that such external aid is necessary. At higher fault currents the more usual series-multiple-arcing mode of fuselink operation takes over and ensures easy circuit interruption.

The UK fuse-switch gear is thus fully self-protecting at all possible fault levels and consequently the fuselinks may be of the back-up variety. Such fuselinks having minimum breaking currents (MBC) in the region of 2·5 to 5·0 times their rated current are suitable. Satisfactory co-ordination may be achieved by making the switches capable of breaking at least seven times the rated current of the largest fuselink used.

The statistical chances of faults occurring below the minimum safe breaking current of the fuse are, in UK practice, quite small in any case, since unearthed neutral points (which could result in small capacitive earth fault currents flowing) are not used and low-voltage secondary-feeder fuses ensure that low-voltage (LV) faults or overloads do not have to be cleared by the high-voltage fuses. Both back-up and general-purpose fuselinks are suitable for such circuits and the latter type has, in the past, been preferred in some applications where instantaneous striker-tripping facilities were not provided because the ratios of their minimum safe breaking currents to minimum fusing currents are lower than those of back-up fuselinks. The usage of general-purpose fuselinks is now reducing, however, because of recent developments which have led to the introduction of high-voltage current-limiting fuselinks capable of clearing all currents between their minimum-fusing and rated-breaking levels (i.e. full-range fuses).

As early as 1965, Mikulecky published a paper entitled 'Current limiting fuse with full-range clearing ability' [35]. The term full-range has recently been included in fuse specification *IEC 60282-1* and it is now widely used by manufacturers and users to describe any HV current-limiting fuse which can, unaided, safely interrupt currents from the rated breaking capacity down to the smallest current which will cause melting of the elements, even under conditions of restricted air circulation. Clearly such fuselinks may be used as the sole protection in circuits operating over a wide range of normal and abnormal conditions.

5.2.3   Full-range fuselinks

Full-range fuselinks of different designs which are now produced by manufacturers, are described below:

(i) Fuselinks with elements operating at high current densities

It was stated earlier that the minimum safe breaking current of a fuselink reduces as the current densities in its individual elements at rated current are raised. In the past, there were constructional difficulties associated with producing fuselinks with large numbers of elements of very small cross-sectional area. More recently, new assembly methods have emerged which have largely overcome such problems.

(ii) Fuselinks in which electronegative gases are produced

The heat produced when arcing takes place in these fuselinks causes large quantities of electronegative gases to be liberated from solid materials within the bodies. The turbulence, cooling and de-ionising effects of the gases cause the arcs to extinguish after many cycles of burnback. In some designs the solid gas-evolving material forms

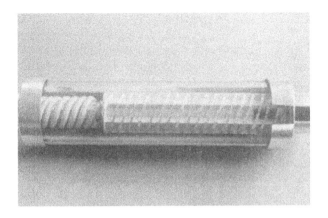

Figure 5.13   Full-range fuse with two series elements

part of the supporting former for the elements and in others it is in the form of plates or beads attached to the elements at points along their lengths.

(iii) Fuselinks with two series elements

Expulsion fuselinks, which are not current-limiting, have been described earlier in Section 5.1.1.

In the past, such fuselinks were comparatively large, but the sizes of the components have now been reduced sufficiently to enable them to be mounted in fuse barrels of standard dimensions in series with current-limiting elements to form integral full-range fuselinks. A typical fuselink of this type is shown in Figure 5.13.

Fault currents above the minimum fusing level up to about five times the rated value are cleared by the expulsion process, the higher currents up to the rated breaking capacity being cleared by the current-limiting elements. The resulting dual-element time–current characteristics closely match the withstand capabilities of protected equipment such as transformers and the co-ordination with other protection devices used on systems is superior to that which can be achieved with many other types of fuses.

(iv) Other types of fuselinks

In recent years several other full-range fuselinks have been marketed or proposed.

In one design, minute explosive charges are placed at points along elements to ensure the formation of breaks simultaneously when the element temperatures exceed the melting-point level.

Full-range fuses of various types are now being produced and installed in many parts of the world. The international standard *IEC 60282-1* has been amended to include special tests for these fuses and some designs have already been tested and approved.

The range of types and ratings available is increasing rapidly, and their costs are beginning to fall towards those of comparable back-up and general-purpose fuses.

5.3   Continental European practice

Only cartridge-type, current-limiting fuses are produced in Europe. They are exclusively used there and in many countries throughout the world to which they are supplied.

The fuselinks are basically similar to the UK designs described earlier in this chapter (Section 5.2) except that the majority do not have elements with low-melting-point metals on them to produce the M-effect. They are designed for compliance with the dimensional standard *DIN 43625* and are usually fitted with cylindrical cap contacts with a standard external diameter of 45 mm and are as illustrated in Figure 5.14. The contacts are often silver plated to ensure that low-resistance connections are obtained when the fuselinks are inserted in their clips. The bodies are produced in four lengths, 192, 292, 442 and 537 mm, being related to the operating voltages and unlike UK practice, are sometimes of organic glass fibre material. Their breaking capacities are similar to the corresponding types produced in Britain, values in the range 30–50 kA being typical. The levels are not specified in any standards and are usually governed by the outputs available in the testing stations.

Operation indicators and/or strikers are optional on all ratings but these are usually spring assisted rather than incorporating an explosive charge to drive them.

An early, and still widely used, arrangement, is to have a wall-mounted switch of an open-frame type with the fuselinks mounted either on the same frame or nearby on a set of separate fuse mounts. This arrangement has the advantage that there is

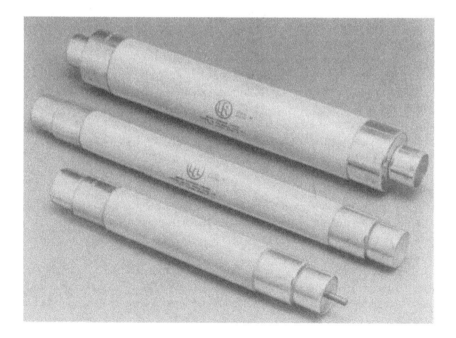

Figure 5.14   Continental European fuselinks

adequate air circulation around the fuses and they are normally able to be used up to their rated current values. The arrangement can, however, be somewhat uneconomical in terms of space and, of course, it is not practical to arrange for striker tripping when the fuses are on separate mounts.

Later, more-compact versions invariably have the fuses mounted integrally with the switch, and striker tripping of all three phases is available as an optional feature. These units are commonly fitted in metal-clad cubicles.

High-voltage fuselinks are also fitted into other types of fuse-switches, including air-insulated ring-main units. Whilst arc extinction is achieved in the switches in several ways, including air-break, hard-gas, oil or, more recently by using sulphur hexafluoride (SF_6), the fuses are invariably in air. The practice of immersing fuses in oil, so widespread in Britain, is almost unknown on the European continent. All-insulated cast-resin fuse-switch units are popular. They do not have metal enclosures and are extremely compact, smaller even than equivalent British oil-insulated designs. The fuses are totally enclosed in cast-resin housings and the resulting severe restriction on heat dissipation usually necessitates appreciable derating of the fuses, i.e. the nominal rated current of the fuses must be higher than the current rating of the unit in which they are installed. A further limitation of this design is that the temperature rises of the exteriors of the fuses must be restricted during operation to eliminate the possibility of damage to the surrounding material of the fuse carriers; 160°C is a typical upper limit. The fuselinks for these units are usually available with strikers to allow three-phase clearance to be obtained if necessary.

As stated above, Continental European fuselinks do not usually take advantage of the M-effect and they therefore tend to run and operate at appreciably higher temperatures than UK designs of corresponding rating. In practice this means that fuses of UK design tend to require less derating than those produced in Europe when mounted in similar enclosures, or alternatively they can safely carry a higher over-current without exceeding permitted temperature-rise limits. Conversely, for a given duty both the rating and physical dimensions of a UK fuselink may often be smaller than those of the European design which would be needed.

In order to obviate the problems due to lack of the M-effect feature, many types of Continental European HV fuses are now fitted with so called 'thermal strikers'. These spring actuated devices are designed to operate and trip associated switchgear when the temperature within the fuse exceeds a safe limit. Such operation may occur BEFORE actual melting of the fuse elements under low overload conditions.

5.4 North American practice

Because of the tremendous demand for electricity and the large number of independent utilities in the USA, HV fuses are much less standardised than they are in other parts of the world. Not only are they produced in many different physical sizes but there are also more basic types to provide the performance characteristics that are required. The distribution systems are also different in that the HV supply is normally single phase, derived from the line and neutral of a three-phase system and the HV supply is brought close to the domestic consumer. Overhead lines feeding small pole-mounted transformers are a common feature in many urban areas; more recently, however, underground residential distribution (URD) has been introduced.

Traditionally, current limiting fuses have been classified as either 'power' or 'distribution' types. The former can be of high current rating, are for use near the power source where TRV values may be high and are for use on three-phase systems. The latter are for less onerous duty on single-phase distribution lines or for protecting small distribution transformers. Highest current rating for these distribution types is normally only 40 A.

The US equivalent of the *IEC 60282* standards is the *ANSI/IEEE C37* suite of standards, see Section 8.4.2. Despite major work on harmonising the two systems, there are still a number of differences between the two documents.

5.4.1   Current-limiting fuses

A significant difference between the constructions of the current-limiting fuselinks employed in the USA and those produced in the UK is that the bodies are often made from organic insulating materials such as resin-impregnated woven glass cloth.

As stated earlier, high-voltage fuselinks may be categorised as either full-range or partial-range (back-up). Both these types are widely used in the USA. The back-up fuselinks are invariably used in conjunction with expulsion fuses, the latter clearing the low-level faults.

High-voltage fuses in North America are mounted in various ways. For pole-mounted transformers, the fuselinks may be mounted directly on the transformer bushings or fuse bases may be provided. For pad-mounted transformers the fuselinks are sometimes fitted inside the transformer tank, inside the transformer bushing, or in oil-tight or 'dry-well' canisters in the side wall of the transformer tank.

The requirement for 'full-range' fuselinks in the USA exists because striker tripping of fuse switches giving very effective full-range performance (widespread in other parts of the world and stemming from UK practice) is little used in the USA. Furthermore, low-voltage feeder fuses either may not be provided or may not be fitted close to the transformer secondary terminals and both these situations could give rise to low-level fault currents flowing in the HV fuselinks.

5.4.2   Non-current-limiting fuses

In the past, low-breaking-capacity, non-current-limiting fuses of various types were very popular and widely used in North America. More recently, however, the increase in fault levels and other factors, such as objections to the noise made by expulsion fuses in clearing, and the move to more underground residential distribution, has caused a swing in demand to the cartridge-type fuses described in the previous section. Nevertheless, non-current-limiting fuses are still being produced and installed in large numbers. One major use is to combine them with back-up fuses which, in this application, are called 'fault limiters'. When a pair of devices is combined in this way, the non-current-limiting fuselink clears the low over-currents, which the back-up fuselink might not interrupt satisfactorily, and of course the latter operates when very high currents flow during short circuits or major faults. Because the fault limiter need only be able to clear currents in a limited range below its rating breaking capacity, it can be produced with a small number of elements of relatively large cross-sectional area. Such constructions are compact and robust, both mechanically and electrically. The overall combination therefore provides economical, full-range protection.

The main types of non-current-limiting fuses which are produced in North America are:

(*a*)   expulsion fuses
(*b*)   'weak links'
(*c*)   fuses containing boric acid.

5.4.2.1   Expulsion fuses

These are very similar to the corresponding UK devices and are used in open cutouts. A large degree of dimensional standardisation has come about in recent years. The majority of types in widespread use are now standardised and mounts and carriers are interchangeable.

The fuses tend to be more compact for a given rating than their UK counterparts because of a trend to single venting designs: a solid metal cap replaces the diaphragm at the top end of the carrier tube and the arcing gasses are reflected downwards so that the total length of tube available for the expulsion process is the same, but overall dimensions and hence also costs are reduced.

5.4.2.2   Weak links

These are basically expulsion fuselinks mounted inside the transformer, the oil quenching the arc during operation. They have, however, a lower breaking capacity than expulsion cutout fuses. Some designs can detect the overloading of the transformer to some extent since their melting currents can be reduced by an increase in oil temperature.

5.4.2.3   Boric acid fuses

These fuselinks are similar to the expulsion fuses described in Section 5.2.2 and they contain springs which ensure that rapid separation and a long break are obtained when melting of the element occurs. This makes it necessary to use special element constructions.

The fuselink body consists of a laminated bakelite tube with a baked-on coating to give it an arc-resistant and weather-proof finish. The body contains a vulcanised-fibre tube within which is a boric acid liner. This forms the chamber in which arc extinction occurs when interruption is taking place, the boric acid providing a de-ionising action which is effective at high current levels.

Boric acid fuselinks are produced for use in three-phase systems at rated line voltages ranging from 23 to 138 kV. They are available with three-phase symmetrical breaking capacities of up to 1000 MV A.

Poles and handling tools are available to enable fuselinks to be removed or replaced after operation. The use of these tools to swing out healthy fuselinks to achieve isolation is inadvisable, and load currents should not be interrupted in this way because arc-control devices are not provided.

Where there is a requirement for HV fuses of very large current rating (say more than 200 A), an alternative to the boric acid fuse now available is the electronically actuated fuse or commuting current-limiter. In these types, the main current carrying conductor is interrupted by explosive charges set off by electronic means. The current is then shunted through a conventional HV current-limiting fuse which ensures that the device as a whole can handle high fault currents and is current-limiting.

Such devices tend to be complex and expensive, so are likely to be found only where more conventional fuses could not be used.

5.4.2.4   Automatic sectionalising link (ASL)

These devices sometimes referred to as 'smart fuses' are in fact not fuses in the sense of operating through fusion of metallic elements. However, since they are designed as retrofit replacements for conventional expulsion fuse carriers and are installed and maintained using fuse operating poles, they are included here as an illustration of how modern technology can affect very marked improvements in overhead line system protection.

The reasons for the development of the ASL were concerned with the discovery that the majority of over-currents on overhead line systems were of a transient, no-damage nature, e.g. lightning strikes, conductor clashing, etc.

Conventional expulsion fuses were operated by such transients and much unnecessary outage and cost resulted. Utilities which did away with expulsion fuses for the above reason were faced with the alternative problem, that when a genuine permanent fault did occur, the upstream switching device would eventually lock out, thereby spreading the resultant outage over an unnecessarily wide area.

The ASL makes use of advances in solid-state microelectronics and is illustrated in Figure 5.15. A miniaturised logic circuit is fitted inside the dimensions of the carrier

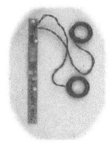

Figure 5.15   Automatic sectionalising link

tube of a conventional expulsion fuse. The tube is a conductor instead of insulator, so the line current flows through this tube and encircling current transformers feed information regarding the state of this current into the logic circuit within.

On the occurrence of a transient over-current, the upstream re-closing circuit breaker will trip and then re-close. The ASL records this event and retains it in its memory for several seconds. If on re-closure the load current has returned to normal, the ASL 'goes back to sleep' and the circuit returns to normal.

In the case of a permanent fault, such as a fallen line or transformer winding fault, when the re-closer, after tripping out, re-closes, the fault will still be present. The ASL logs this second event and awaits the second trip-out of the re-closer. When the ASL detects that line current has fallen to zero, it fires a chemical actuator (similar to the fuse striker described in Section 5.2.1). This de-latches the ASL carrier tube so that it swings down and provides circuit interruption and isolation distance.

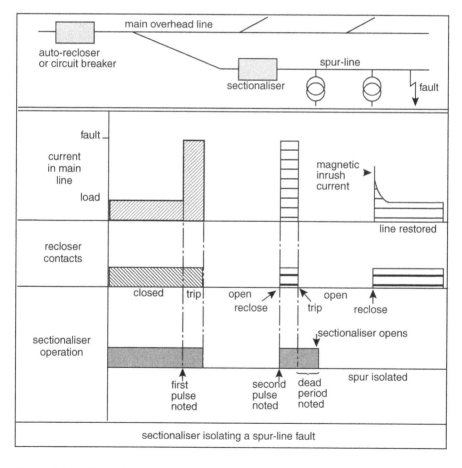

Figure 5.16   Typical operation of ASL (sectionaliser)

When the re-closer closes for the second time, the faulty section of line has now been disconnected, so the remainder of the network can revert to normal operation.

The linesman brings down the operated ASL by means of a conventional fuse operating pole, fits a new actuator and lifts the ASL back into position.

In some versions of the ASL, a re-settable magnetic latch is used in place of the chemical actuator. The latch obviates the need to fit a replacement actuator but is more complex and costly.

The operation of an ASL in a typical UK distribution system is illustrated in Figure 5.16. The sectionaliser referred to in Figure 5.16 is the automatic sectionalising link, ASL.

A very large number of ASLs are now in use around the UK and have proven their value in terms of reduced maintenance costs and reduction in customer hours lost. The electronic ASL was an exclusively British invention and development and is now also in use in many countries overseas.

Chapter 6

Constructions of miniature, plug and other small fuses

Although miniature, domestic plug and other small fuses are physically similar, they are grouped into several different application categories and must comply with different specification standards.

Miniature fuses are defined as being for the protection of electric appliances, electronic equipment and component parts thereof, normally intended for use indoors. A British Standard, *BS 646*, for such fuses, was introduced in 1935, but it was not until after World War II that they became of technical and economic importance because of the rapid development of the electronics industry.

Plug fuselinks are classified as domestic fuses, which are defined as being for use in domestic and similar buildings, for example dwelling houses, blocks of flats and office buildings. Within this class are the semi-enclosed and cartridge fuselinks used in equipment such as supply-authority fuses and consumer units. The constructions of these fuses were described in Chapter 4 because of their similarity to other low-voltage fuses.

This chapter will deal with the smallest types of fuselinks that are produced, namely the miniature and domestic plug fuses referred to above and also automotive fuses. The total world market for these components is estimated at being well over 10 billion per annum and the continuing increase in the number of small electrical appliances being marketed makes it essential that the designs and methods of manufacturing these fuses should enable them to be produced in large quantities at low cost with high quality and reliability. The various types of fuses differ in some important respects and therefore they are considered separately in the following sections of this chapter.

6.1 Miniature fuses

Many different types are available with current ratings from 32 mA to 20 A, a selection being shown in Figure 6.1. Most of them have cylindrical bodies and ferrule-type

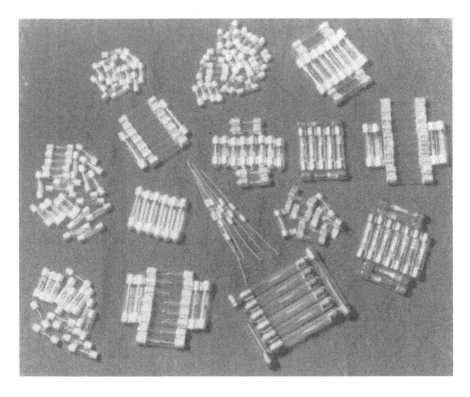

Figure 6.1   Selection of miniature fuselinks

end caps, though an increasing number are produced for ease of fitting into printed circuit boards. The IEC standard covering miniature fuses is *IEC 60127, Part 1* of which covers definitions and general requirements. The three main types of miniature fuselinks are covered by *Parts 2, 3* and *4*:

Part 2   Cartridge fuselinks
Part 3   Subminiature fuselinks
Part 4   Universal modular fuselinks

6.1.1   Cartridge fuselinks

IEC Specification 60127, Part 2 only covers fuselinks with rated currents up to 6·3 A and dimensions of 5 mm diameter and 20 mm in length. These are the most widely used cartridge fuselinks, followed by a slightly larger design of 6·3 mm by 32 mm.

Miniature fuses are categorised according to their breaking capacities and speeds of operation. Those in the high-breaking-capacity category must be able to interrupt alternating currents of 1500 A at voltages up to 250 V. They are generally manufactured with ceramic bodies filled with granular quartz. Low-breaking-capacity fuselinks must be able to interrupt alternating currents up to 35 A or ten times their rated currents, whichever is the higher, at voltages up to 250 V. These fuselinks usually have glass bodies and no filling material and as a result their elements can readily be inspected. The breaking classification is marked on the end cap of each fuselink, the letter H indicating high-breaking-capacity and L indicating low-breaking-capacity.

A further breaking capacity category, E, of 150 A was introduced because the 35 A rating is not adequate for many modern applications and indeed several authorities now prohibit the use of low-breaking-capacity fuses.

The categories of speed of operation are also signified by internationally accepted letters that are marked on the fuselinks. The letters and corresponding categories are as follows:

FF	Super-quick-acting
F	Quick-acting
M	Medium-time-lag
T	Time-lag or anti-surge
TT	Super-time-lag

Figure 6.2 gives an indication of the different operating speeds of the various categories.

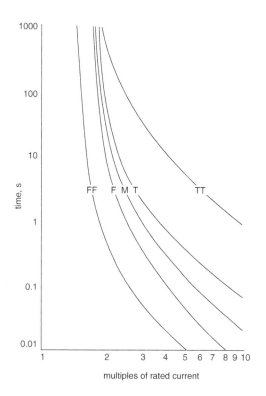

Figure 6.2   Time/current characteristics of miniature fuselinks

Quick-acting fuselinks (category F) invariably have single wire elements. In the past, most of these elements were of silver, but the high cost of the material caused manufacturers to use other materials where possible and now silver is only used in fuselinks with relatively high current ratings. Silver-plated copper wires are now used widely in fuselinks rated between 1 and 5 A. Copper-alloy wires are used in fuselinks rated below 1 A and at the lowest ratings nickel and nickel–chromium wires are used. These latter materials which have high resistivities are preferable in fuselinks with very low current ratings because they permit larger diameter wires to be used for a given power loss than would be possible with wires of low resistivity. This is a very important factor affecting manufacture and reliability, as can be appreciated by considering a 50 mA fuse. Such a fuselink might well have a nickel–chromium wire element of 0·013 mm diameter and such fine wire clearly needs to be handled with great care during production. If silver or copper elements were to be used they would have to be of even smaller diameters, and as the tensile strengths are lower than that of nickel–chromium, the elements would be almost impossible to handle and the completed fuselinks would be unacceptably fragile. These same considerations have also led to fuselinks of low current ratings being produced without filling materials around the elements, because the presence of such materials would improve the cooling and thus necessitate finer wire elements to give the same running temperatures at the rated current levels.

Fuselinks with very low current ratings have quite high resistances; for example the 50 mA fuselink, to which reference has been made above, would have a resistance in the region of 150 Ω and, because of this, such fuses are current limiting. In any case, they are only required to interrupt low currents and so the need for the improved arc-quenching performance associated with filling material is not necessary. For the same reason it is usually possible to employ tubular glass bodies which are not only cheaper than those made of ceramic but have the advantage that the element is visible and thus it is possible to see whether operation has occurred.

It must be recognised that voltage drops occur across fuselinks when they carry their rated currents and these could be unacceptably high when used with equipment operating at a low voltage. As an example, a fuselink rated at 250 V would probably be unsuitable for use in a circuit operating at 6 V. For such an application, a fuselink with a very short element and thus relatively low resistance, should be used, to limit the voltage drop across it at rated current, or a higher current rating could be used.

Fuselinks with higher current ratings which are to be used in circuits where fault currents may exceed 35 A are normally filled with quartz of controlled grain size and tubular ceramic bodies are used.

Super-quick-acting fuselinks (category FF) are generally similar to category F fuses except that the elements have a restricted section in them, i.e. a portion of reduced cross-sectional area. These elements are produced by plating the end sections of the wire to increase the area or etching material away at the central sections.

This non-uniformity produces the same effect as the restrictions in the elements of high-voltage and low-voltage fuselinks, namely the energy let-through ($\int_0^t i^2 \, dt$) to the protected device is limited. These fuses are mainly used to protect components, such as relatively small semiconductor devices, which are very sensitive to

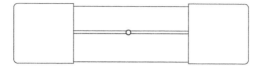

Figure 6.3   Fuselink containing a tin-alloy pellet

over-currents. Such fuselinks are expensive to produce and the demand for them is small, because it is more cost-effective to use semiconductor devices with current ratings somewhat above the actual circuit rating, so that they can withstand the currents let through by simpler and lower-cost category F fuses.

Within the time-lag categories (M, T and TT), the following three types of fuselinks are in general use. They are mostly of low current-breaking capacity with glass bodies and no filling material. Versions with ceramic or Pyrex tubular bodies containing granular quartz filling material are now being produced, however, with breaking capacities up to 1500 A at 250 V AC, and they are being widely applied to electronic equipment where there is now increasing stress on safety.

The first type employs elements made up of two materials to enable the performance associated with the M-effect to be obtained. Three different forms of element are used. One form employs one or more so-called diffusion pills and the other two use composite wires.

In the first form one or more small pellets of tin alloy, referred to as diffusion pills, are soldered or lugged on to a silver or silver-alloy wire along its length, an example being shown in Figure 6.3. In these fuses, diffusion occurs at the interfaces between the pellets and the wires as element temperatures rise during operation and when the melting-point temperature of the tin alloy is approached, a region of high resistance is created and ruptures then occur at the positions of the pellets.

As stated earlier, composite wires of two forms are used as elements. In one form in which a clad wire is used, the higher-melting-point material (silver) surrounds a core of low-melting-point tin alloy whereas in the other form, which uses plated wires, the low-melting-point material surrounds a silver or silver–copper-alloy core. This latter form can be regarded as the limiting situation of diffusion pills in which one pill extends along the whole length of the element.

All three forms of elements are used in fuselinks with current ratings in the range 800 mA–10 A and at ten times rated current their minimum pre-arcing times are 20 ms.

They are thus able to withstand short, isolated current surges and they provide the delays required in the medium-time-lag category (M). Their surge resistance is such that they will not operate when subjected to repeated transient currents of ten times the rated value, and 10 ms duration, provided that the interval between successive transients is at least 30 s. This interval is, of course, required to allow the element to cool between the current pulses.

Clearly, the production of composite wires of very small diameters is fairly costly but this factor is offset by the simple production of fuselinks using elements of this type.

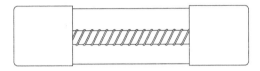

Figure 6.4   Fuselink with a helically wound element

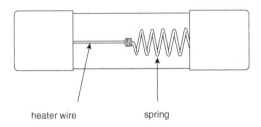

heater wire   spring

Figure 6.5   Fuselink containing a coil spring

The second type of time-lag fuselink, which is illustrated in Figure 6.4, incorporates a helically wound element. In this construction a silver or copper wire is wound on a stranded, insulating former and the pitch of the turns is maintained within close limits to ensure consistent performance. Because of the length of the wire in this type of element, it must be of greater diameter than that of a straight-wire element, to give a comparable resistance. The resulting mass of the element material causes the rates of temperature rise to be low and the operation is relatively slow. Category T and TT performance is thus readily obtainable.

These fuselinks are made in ratings from 25 mA to 10 A and they offer good surge resistance. They do not make use of the M-effect and they are thus not greatly affected by ambient temperatures, because of the relatively high melting temperatures of the element materials.

The third type of fuselink construction used to obtain time-lag operation is shown in Figure 6.5. The element consists of a coil spring which is soldered with a low-melting-point alloy to a wire, usually referred to as the heater wire. The joint is usually located midway along the fuselink. The springs and heater wires are typically made from copper alloys, the spring materials being hard drawn.

At high overcurrents, the heater wire ruptures before the joint melts and the spring contracts into the end cap, good clearance thus being effected. Care must be taken in the design stage to ensure that the spring will not anneal before the heater wire ruptures, because otherwise the performance after the passage of a short pulse of high current would be affected.

At lower over-currents, the joint melts and the spring again contracts into the end cap ensuring good clearance. The operating times for these conditions depend on the specific heats, masses and physical dimensions of the spring and solder, because these parameters control the heat-transfer processes, and it is possible to obtain the time lags required of category T and TT fuselinks. Pre-arcing times in excess of 50 or even

100 ms are readily achievable with this construction at currents of ten times the rated values, which range from 32 mA to 15 A.

To prevent the spring from overheating a bypass wire is connected across it in fuselinks with rated currents above 1 A.

It is necessary to use solders with relatively low melting points in these fuselinks and as a result their performances may be significantly affected by the ambient temperature. Certainly the current ratings must be reduced if temperatures above 50°C may be encountered.

Fuselinks of this type were popular in the past but in recent years their usage has reduced considerably.

All miniature fuselinks, as stated earlier, have ferrule-type end caps, usually made of copper or brass suitable for deep drawing. They are silver, nickel or tin plated to ensure that good connections are obtained when they are mounted in their holders.

Some designs have a single cap at each end whereas others are doubly capped. Fuselinks with glass bodies are usually single capped. In one construction, solder in each end cap is used to form the joint between the element and the end cap, and adhesion between each cap and the body is effected by the solder and the flux in it. In an alternative construction, which tends to be used at the lower current ratings, the end caps are glued to the body and then the element is threaded through holes in the end caps using a needle, after which connections are made by soldering externally. This latter method is now being used less frequently because solder contains lead and there is thus a possibility that a person handling such fuselinks could suffer because of the presence of this toxic substance.

Fuselinks with high breaking capacities and ceramic bodies generally have double caps, the element being connected to the inner caps. For the lower-breaking-capacity types with ceramic bodies single caps are often used.

Unlike other types of fuses, there are no significant constructional differences between the miniature fuselinks used in the UK, Continental Europe, North America and Japan. They are, however, produced in different physical sizes.

In Britain, the predominant type was 32 mm long by 6·3 mm diameter but the majority are now being made 20 mm in length and 5 mm in diameter. The fuselinks manufactured and used in Europe are normally of these latter dimensions.

In North America, miniature fuselinks are produced to comply with specification *UL 248-14* which states performance requirements in detail, but does not include standard dimensions, leaving them to be determined by manufacturers' traditions or the preferences of users. The most common size is 32 mm long by 6·3 mm diameter, but 20 mm in length and 5 mm in diameter are becoming more common. The fusing factor specified in *UL 248-14* is lower than that required in *IEC 60127*. To achieve the lower fusing-factor and rated-current levels, elements of smaller cross-sectional areas are used for given current ratings than is the practice in the UK and Europe.

A further factor which has also led to differences between the standards in Europe and North America is that the domestic supply voltage in the latter area is only 115 V and this makes it easier to clear fault currents. Because of the above differences in definitions, it must be recognised that fuselinks produced to meet the requirements of *UL 248-14* are not interchangeable with those produced to satisfy *IEC 60127*.

In the 1990s, a smaller size of 15 mm length and 5 mm diameter has been introduced, particularly in North America and Asia Pacific. The reduction in size meeting the need for more compact equipment.

The selection of the appropriate miniature fuselink for a specific application is based on rated current and the ability of the fuselink to withstand normal circuit overloads, i.e. which type of time/current characteristic is appropriate. Specific references to miniature fuselinks will therefore not be found in Chapter 7 which deals with applications.

A wide variety of methods are used for mounting the fuses in the equipment including:

- panel-mounting fuse-holders
- 'in-line' fuse-holders
- base-mounting chassis type
- printed-circuit-board mounting fuseholders, bases, clips and leaded fuselinks.

Traditionally the input fuselinks protecting the whole equipment were panel mounted for replacement of the fuselinks on the outside of the equipment. A typical panel-mounting fuse-holder is shown in Figure 6.6. Panel-mounted fuseholders usually fit into a 'D'-shaped hole in the rear or front of appliances. This D-shape gives the correct orientation and stops rotation. They consist of a cylindrical base fixed in the panel with a front or back nut or clip. The fuse is fitted into a carrier, which is screwed into the base by finger grip or through a screwdriver slot or bayonet fitting. The carrier allows safe fuse replacement although the power should be switched off first. Some designs incorporate a neon lamp, housed in the fuse cap, to provide an indication in the event of the fuselink operating and the supply remaining connected. Cable connections are made onto terminals protruding from the base by push-on 'fast-on' or 'solder' tags.

There are therefore a large number of panel-mounted fuse-holders to suit the needs of equipment manufacturers.

An increasingly important category of enclosed fuseholder is the so-called shocksafe, which by design prevents the user coming into contact with live parts.

Enclosed fuseholders must be capable of dissipating the power transferred to them from the fuselinks without causing unacceptable temperature rises. The IEC is moving towards specifying the maximum power dissipation that a fuseholder can safely endure. Three categories are currently specified: 1·6, 2·5 and 4 W. In the UL system, fuseholders are rated according to the maximum currents that they can carry. Experience shows, however, that considerable derating should be applied and in general fuseholders should not be run at more than 50 per cent of their UL ratings.

It is important, particularly at the higher current ratings, that low-resistance connections should be obtained between the fuselinks and the contacts in their holders, and therefore silver-plated, copper-bearing spring materials are usually employed to make the contacts in the holder.

Today the fuses are normally changed by a service engineer, rather than an unqualified person, since the nature of the fault has to be identified in complex circuits. If the fault is still present, it will merely operate the new fuse that has been fitted. In view

Figure 6.6   Fuse-holders

of the above, there has been a decline in the use of panel-mounted 'mains' fuses to the more economical solutions as follows.

In-line fuse-holders are normally enclosed and have flying leads or sometimes push-on connectors. Simple open-type fuse bases for mounting onto chassis are another option and are also shown in Figure 6.6.

A more common approach is fitting into through-hole printed-circuit-boards and a number of options are available including:

- enclosed fuse-holders in either vertical or horizontal orientation
- open-type fuse bases, with covers as an option
- fuse clips, suitable for automatic insertion
- leaded fuselinks suitable for automatic insertion.

6.1.2   Subminiature fuses

Miniature fuses tend to be too large for many printed-circuit-board applications and therefore the subminiature class was introduced. It is specified in *IEC 60127 Part 3* that the bodies of such fuselinks should have no principal dimension greater than 10 mm.

These fuselinks are produced in 125 and 250 V ratings with cylindrical polymer bodies and two leads protruding radially from them. They are also produced with axial leads, the voltage rating for these fuselinks being 125 V. Both types are also covered by *UL 248-14*.

Whilst holders are available for these fuselinks, it is the usual practice for them to be soldered directly to printed-circuit-boards.

Fuselinks are produced in the five operating-time categories referred to in Section 6.1, namely FF, F, M, T and TT, the types F and T being used for the majority of applications. As with miniature fuselinks, simple wire elements are used to obtain category F performance and composite elements to produce the M effect are used to obtain the time delays required by category T. These fuselinks are only available with low breaking capacities up to 1500 A.

As fuselinks are made smaller, it becomes increasingly difficult to obtain the required breaking capacities at 250 V ratings because of the limited arc lengths that can be achieved. Various techniques have been used to obtain desired performances. In one design the interiors of the fuselink bodies are divided into several chambers and in another use is made of the ablative effect, gas given off from the plastics body in the fuselink during arcing assisting extinction. Such fuselinks are relatively expensive and, in many cases, it is economic to design electronic circuits to operate at lower voltages, such as 125 V, so that they can be protected by simpler fuselinks.

6.1.3   Universal modular fuselinks

A range of even smaller fuses designated as universal modular fuses (UMF) has been proposed. These fuselinks would be available in voltage ratings of 32–250 V and the following operating-time categories, the times being those at ten times the rated current:

FF	not exceeding 0·001 s
F	0·001–0·01 s
T	0·01–0·1 s
TT	0·1–1 s.

All categories must operate in 10 s at twice the rated current and be able to carry 1·7 times the rated current for 1 h. The techniques used in miniature fuselinks would be used in UMF devices to obtain the various operating-time characteristics.

These fuselinks would be produced in two forms: one, designated SMT, would be suitable for surface-mounting connection, the leads being soldered directly to the printed-circuit-tracks; the leads in the other form would be taken through holes in the printed-circuit-boards. The latter is described as through-hole mounting (THT).

Fuselinks with different voltage ratings have different spacings between their connecting leads to make them physically non-interchangeable.

6.1.3.1   'Chip fuses'

As consumer electronic products become smaller and more portable, printed-circuit-boards have become more densely populated, forcing fuses located on the board to become even smaller. Additionally, more stringent electrical performance criteria have been imposed on these types of board level fuses.

Cost, size and performance are chief aspects considered by the end user. Low cost is by far the most important criteria. Next is the component size. Both the amount of space that the component occupies on the printed-circuit-board and a low profile are highly desired, particularly in slim line, hand-held type consumer electronic products.

But, performance cannot be overlooked. The electrical characteristics of the fuse are sometimes compromised by cost and size. Desired electrical characteristics include: high pre-arcing I^2t, to be able to withstand inrush currents; low resistance, so that high voltage drop is not created across the fuse; good current and temperature cycling capabilities; and high breaking capacity.

Board level fuses, or more commonly called 'chip fuses', get their name because the construction is very similar to that of chip resistors and chip capacitors. The fabrication techniques and manufacturing methods are similar as well.

Materials, construction techniques and manufacturing methods are borrowed from the electronics industry. Some fuse designs are based on hybrid microelectronic technologies, while other designs borrow techniques and methods from the printed-circuit-board industry. Hybrid microelectronic technologies include thick and thin metal films on glass and ceramic substrates. Printed-circuit-board technologies comprise copper foil or plated copper on a polymer-type substrate.

Designs based on hybrid microelectronic technology are very similar to the common chip resistor. But, instead of a resistive element, a fuse element is deposited on to a ceramic or glass coated ceramic substrate. The metal film deposition may be achieved by screen printing, vacuum vapour deposition or the like. Additional processing techniques such as laser trimming may be incorporated as well. The fusing element can be laser trimmed to a specified resistance, much like resistor trimming.

Various types of ceramic substrates may be utilised in the design. 'As-fired' or green tape type ceramics are used. These types range from the common alumina (Al_2O_3) or any compound within the $Al_2O_3 * MgO * SiO_2$ ternary system. The type of ceramic substrate selected will affect the electrical properties of the fuse based on the arc quenching capability, thermal conductivity and dielectric properties of the ceramic substrate. The substrate in combination with the fusing element material, element geometry, and the material used to encapsulate the fuse element all contribute to the overall fuse characteristics.

In some cases, an arc quenching media is used to surround the fusing element and contain the arc. Ideally, the material will have a high arc track resistance, will not carbonise at the fusing temperature and it will have good mechanical strength. Preferably the material will change phase and absorb some of the heat of fusion that is created when the fuse element vaporises and opens the circuit. This phase change may be a solid glass into molten glass, silica into glass, or silicone into fumed silica. Ceramic, glass, silicone and composites of these materials are known to have good arc-suppressing characteristics.

Multiple fusing elements also may be used to enhance fusing characteristics, in applications where high current and/or high voltage is needed. Two or more fuse elements may be connected in series or in parallel. Depending on the desired voltage and current carrying capacity of the fuse. This type of design is very similar to the construction of a multilayer chip capacitor. The fuse elements on the substrate are vertically stacked. Parallel elements make contact at the element ends and termination interface. Metallised vias through the substrate are used to connect series elements.

Figure 6.7   Chip fuses with M-effect

Designs based on printed-circuit-board technology are constructed of metal foils or plated metallisation on a type of FR-4, glass-filled epoxy material. This type of polymer-based design with its low temperature processing allows the designer to incorporate an M-spot. The M-spot design benefits due to its low resistance and higher pre-arcing I^2t but other characteristics such a current cycling capability and low overload characteristics are compromised. Figure 6.7 with top, bottom and side views illustrates this construction with M-effect.

As in any design, performance trade-offs must be considered. For example, fuses that utilise a ceramic substrate typically have better current cycling capabilities and thermal withstand capabilities than fuses that utilise printed-circuit-board type technologies. But, higher pre-arcing I^2t capabilities are sacrificed.

This is because a thick or thin film fuse element is in intimate contact with the surface of the ceramic substrate and the coefficient of thermal expansion is matched, therefore there is minimal mechanical stress due to thermal expansion and contraction while the fuse element heats and cools during current cycling. On the other hand, fuses that utilise printed-circuit-board type techniques are fabricated using lower temperature processing techniques where an M-spot is used in the design. An M-spot increases the pre-arcing I^2t, inrush withstand capabilities; but current cycling and thermal withstand performance is compromised. This is because the tin M-spot begins to alloy with the base copper element when the fuse heats up during current cycling. As the tin M-spot alloys then the resistance will begin to rise, ultimately causing the fuse to operate.

Very precise resistance that is achieved by laser trimming will also reduce pre-arcing I^2t withstand capabilities. Because the pre-arcing I^2t is a very fast event, within milliseconds, there is no time for heat to be dissipated. Therefore the pre-arcing I^2t is dependent on the current density; the cross-sectional area of the fusing element. Looking at most laser trimmed devices, the element metallisation is relatively thin and the effective fuse element cross-sectional area is small. This provides for a relatively low pre-arcing I^2t. A typical fuse element which is laser trimmed is shown in Figure 6.8.

The highly desired, smaller, low profile fuse has a lower breaking capacity than a larger package size. This is due to two reasons; one, the distance between the electrodes is directly proportional to the voltage withstand and two the mechanical strength of the overall fuse body is reduced.

Figure 6.8   Laser-trimmed fuse element

Table 6.1   Sizes of some chip fuses

Component identifier, inch (mm)	Package dimensions (mm)
1206 (3216)	3·2 × 1·6
0805 (2012)	2·0 × 1·25
0603 (1608)	1·6 × 0·8
0402 (1005)	1·0 × 0·5
0201 (0603)	0·6 × 0·3

Figure 6.9   Standard sizes of chip fuses

Package configurations must be compatible with standard pick and place machines used in the printed-circuit-board industry. Therefore, standard package mechanical outlines and sizes are taken from the Institute for Interconnecting and Packaging Electronic Circuits (IPC) and Electronic Industries Association (EIA) industry standards. These standards are developed by the Surface Mount Land Patterns Subcommittee of the Printed Circuit Board Design Committee of the IPC. The standard provides information on land pattern geometries used for the surface attachment of electronic components. It also provides the appropriate size, shape and tolerance of surface mount land patterns to insure sufficient area for the appropriate solder fillet and also to allow for inspection and testing of those solder joints.

Sizes of such chip fuses are given in Table 6.1 and are illustrated in Figure 6.9.

6.2   Domestic plug fuses

Socket and fused-plug outlets were certainly available in Britain in the early 1930s, and *BS 646*, which included requirements for cartridge fuselinks up to 5 A rating to be used in plugs, was published in 1935. These fuselinks, which were a ferrule design, 19·05 mm long and 5·33 mm diameter, had a breaking capacity of 1 kA. They are still produced, but they have only a limited use in small appliances, such as electric clocks.

The use of 'ring-main' circuits to feed socket outlets in domestic and similar buildings was advocated by a committee of the Institution of Electrical Engineers in 1942. It recommended that a new type of 3 kW socket outlet and fused plug should be introduced. This recommendation was not implemented immediately because of the more pressing needs arising from World War II, but the production of fused plugs, complying with *BS 1363*, was eventually commenced in 1947 and vast numbers of them have been produced and used in 240 V AC circuits since that time. The fuselinks, which are very familiar to most people in the UK, are of the ferrule type, 254 mm long and 6·35 mm diameter. They are available in a range of current ratings up to 13 A and comply with the requirements of *BS 1362*. The two preferred ratings of 3 and 13 A also conform with *IEC 60269-3-1 Section IV*. Initially they had a breaking capacity of 1 kA but in 1953 this was increased to 6 kA at lagging power factors down to 0·3. To enable the fuselinks to achieve this breaking capacity, they are filled with quartz of controlled grain size and have ceramic bodies. The elements are of silver-wire and tin pellets are usually added to them, to obtain the performance associated with M effect. This enables the fuselinks to meet the requirements that their power losses at rated current should not exceed 1 W and that their time/current characteristics lie within specified boundaries which permit the passage of normal overload transients but nevertheless provide a high degree of over-current protection.

The ends of the elements usually pass through inward-facing conical eyelets, or are soldered to washers. Caps of copper or brass suitable for deep drawing are then pressed on, either making a pressure contact with the washers or compressing the ends of the elements between the inner faces of the cones and conical protrusions in the bases of the caps.

6.3   Automotive fuses

In recent years, the amount of electrical and electronic equipment installed in motor vehicles has increased greatly, resulting in more complex 12 V systems and load requirements ranging from 1000 W to in excess of 2000 W in highly appointed vehicles. For circuit protection requirements, a modern middle-of-the-range model can have about 30 fuses in it, and there may be 75 or more in a luxury car.

For many years, simple cylindrical glass fuses were used in the UK and USA and exposed fuse element, semi-cylindrical fuses (commonly called 'torpedo' fuses) were used on the continent of Europe. The current ratings of these fuselinks were not universally standardised and the UK previously rated its fuselinks at the current at which the fuse would operate, the true continuous rating being about half that. In some instances fuselinks were marked with a dual rating; as an example, a marking of 10/20 A on a fuselink indicated that it could carry 10 A continuously and that it would operate at 20 A.

The prevailing nominal voltage in motor vehicles is 12 V and larger vehicles such as trucks, buses, emergency equipment or specialised platforms (e.g., military split systems of 12 and 24 V). As these applications evolved, it became necessary to use fuses that could accommodate either voltage range. A common value of 32 V emerged, as this would handle the highest permissible transient voltage in a system operating nominally at 24 V. For short-circuit ratings, 1000 A became an accepted standard for 32 V fuselinks and more recently, some fuselink designs have capacities up to 2000 A, partly in response to application requirements in systems outfitted with high-wattage components and supplies.

The trend towards greater complexity and electrically operated accessories in vehicular systems coupled with governmental regulations concerning energy conservation and reductions in toxic emissions has driven the transportation industry's research and development towards new solutions. Electrically, this has meant new approaches to supply and overall vehicular electrical architecture design. At present, there are several solutions that have gained some prominence: 42 V electrical systems (or 42 V/12 V split system, a mild-hybrid), high-voltage drive system for electric propulsion motors (such as 300 V) with either 42 V or 42 V/12 V subsystems for conventional components and accessories (a hybrid), and the latest thrust to development of fuel cell supplied systems with a high-voltage AC or DC controller for totally electric propulsion and step-down conversion for all other necessary components. The hybrid-type systems rely on a reduced form of internal combustion engine for supplemental propulsion and system electrical charging while the fuel cell approach uses hydrogen as fuel processed through an exchange catalyst that produces electric current with the 'waste' product being water. All of these newer approaches change the complexity of over-current protection requirements and what sort of over-current device will be required to accomplish protection.

As electrical system voltages and current capacities have evolved, so have the approaches towards fusing. The first major evolution in fuse design occurred in the late 1970s. The cylindrical glass fuses and 'torpedo' fuses, while satisfactory in performance, were not well suited to amperage ratings exceeding 30 A, were not conducive towards reductions in size requirements for mating fuse holders, and were not easily adapted to high-volume vehicle manufacturers desiring to automate the assembly processes of electrical centres in vehicles. In response to this, a blade-type fuse format was developed and introduced. Three basic sizes have emerged in succession as illustrated in Figure 6.10, which have become the dominant type of fuse used in the transportation industry. These fuses, which are available in standard ratings from 1 to 100 A with 32 V maximum, are capable of withstanding the inrush currents that occur in vehicular electrical systems. The two smaller sized blade-type fuses are typically considered as fast acting. The large size blade-type fuse is generally termed as 'time lag' in operation, which means that overloads under 150 per cent of rating take a longer time to operate the fuse than the fast acting styles. This makes the

Figure 6.10   Blade-type automotive fuses

Figure 6.11   Bolt-in automotive fuses

fuses better suited for higher inrush conditions or for short-circuit cable protection. The blade-type fuses utilise plastic housings that are moulded in various colours to allow for a system of visual recognition tying colour to amperage rating, which establishes standardisation for users.

With larger amperage requirements in some components and the need for main fusing, other styles of fuses have emerged, such as bolt-in fuses. These fuses, as the term implies, are installed into holders with threaded mounting posts that are used to affix wire terminals and fuse terminals together (see Figure 6.11). Bolt-in style fuses can cover current ranges from as low as 30 A to as high as 500 A, also with 32 V maximum.

The emerging voltage increases for advanced vehicular electrical systems have resulted in the availability of some compatible fuses. For 42 V system applications, several versions of blade-type fuses and bolt-in fuses have been developed with 58 V ratings, which reflects the anticipated maximum transient voltage in a 42 V nominal system. As with earlier 32 V devices, proper mechanical design and clearances have provided short-circuit performance in either the 1000 or 2000 A levels as unfilled fuselinks. The higher arc potential can be successfully managed in the melting element sections of the fuses. The higher operating system voltages do, however, make it necessary to incorporate certain safety features to prevent users from disengaging or engaging fuses or other pluggable devices in live circuits due to the destructive nature of arc energy on mating terminals.

While conventional fuselink design has application in both older and newer systems, other means of fusing (over-current protection) are being explored. Examples would be: positive temperature coefficient (PTC) devices, see Section 1.3.3, which are solid state in nature and can 'switch' between conducting and resistive states, surface mount devices (SMD) that could be fuses or solid-state devices directly installed on circuit boards used in vehicular systems, electronically controlled power distribution systems that monitor circuit loads and switch on/off or re-route loads, pyro-fuses that incorporate conventional fuselink design but have firing circuits that force a fuse operation in a catastrophic event such as a vehicular crash to cut all power from the supply, and other electronically activated devices such as a Hall-effect sensor in combination with a field effect transistor (FET) that would shut down a circuit when current overload biases the sensor output to the FET.

Remarkably, glass fuses and 'torpedo' fuses are still in use today though predominantly as either service parts or in aftermarket accessory modification products. The blade-type fuse is still widely used at the vehicular manufacturer level due to its wide availability and economy. Alternative protection devices will continue to emerge in response to newer overall technology improvements and economies of scale. Finally, legislative efforts on toxic materials and vehicle end-of-life disposal requirements will affect fuse design and use.

Chapter 7

Application of fuses

7.1 General aims and considerations

Fuses are used for so many different applications that it is impossible to consider all of them and therefore only some of the more common are discussed in this chapter. There are, however, some general aims and considerations which apply in all applications and these are dealt with initially.

A fuselink which is to protect a particular piece of equipment or circuit should ideally satisfy a number of criteria. This is illustrated by considering a simple example based on the circuit shown in Figure 7.1.

First, the minimum fusing current of the fuse should be slightly below the current which the cables and item of equipment are able to carry continuously.

The item of equipment will usually be able to carry overload currents for limited periods, and the fuse should operate at these current levels in times slightly shorter than the corresponding equipment time ratings. Clearly, the cables should also be able to cope with this duty without suffering damage.

Higher currents may flow as a result of faults within the item of equipment and in these circumstances the primary requirement is that consequential damage to the remainder of the circuit should be prevented. The extreme case will occur in the event of a short circuit between the line and neutral terminals of the equipment. Clearance must then be effected before damage is caused to the cables.

A further possibility is a short circuit between the conductors of the connecting cable. The most severe situation would arise if the fault was at the input end,

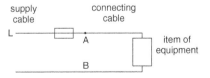

Figure 7.1   Circuit protected by a fuse

i.e. between points A and B, and in these circumstances the fuse would have to interrupt the circuit before the source and supply cables could suffer damage.

This chapter covers typical applications and additional information pertaining to low-voltage, high-voltage and miniature fuses is given in Section 8.1.8 – Application guides in standards.

IEC TR 61818, an application guide for low-voltage fuses, gives a summary of the advantages of current limiting fuses and it is felt appropriate to draw readers attention to these benefits. Many of these benefits also apply to high-voltage and miniature fuses and in particular current-limiting fuses.

The current-limiting fuse provides complete protection against the effects of over-currents by protecting both electric circuits and their components. Fuses offer a combination of exceptional features, for example:

- High breaking capacity (current interrupting rating).
- No need for complex short-circuit calculations.
- Easy and inexpensive system expansion where increased fault currents are concerned.
- Mandatory fault elimination before resetting. Unlike other short-circuit protective devices (SCPD), fuses cannot be reset, thus forcing the user to identify and correct the overcurrent condition before re-energising the circuit.
- Reliability: no moving parts to wear out or become contaminated by dust, oil or corrosion. Fuse replacement ensures protection is restored to its original state of integrity.
- Cost-effective protection: compact size offers low-cost over-current protection at high short-circuit levels.
- No damage for type 2 protection according to *IEC 60947-4-1* and *IEC 60947-4-2*. By limiting short-circuit energy and peak currents to extremely low levels, fuses are particularly suitable for type 2 protection without damage to components in motor circuits.
- Safe, silent operation: no emission of gas, flames, arcs or other materials when clearing the highest levels of short-circuit currents. In addition, the speed of operation at high short-circuit currents significantly limits the arc flash hazard at the fault location.
- Easy co-ordination: standardisation of fuse characteristics and a high degree of current limitation ensure effective co-ordination and discrimination between fuses and other devices.
- Standardised performance: fuselinks according to *IEC 60269* ensure availability of replacements with standardised characteristics throughout the world.
- Improved power supply quality: current limiting fuses interrupt high fault currents in a few milliseconds, minimising dips in system supply voltage.
- Tamperproof: once installed, fuses cannot be modified or adjusted in order to change their level of performance, thus malfunction is avoided.

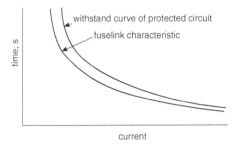

Figure 7.2   Time/current characteristics of circuit and fuse

7.1.1   Time/current relationships

To achieve the criteria listed above, fuselinks should have time/current characteristics which lie close to the withstand curves of their associated circuits as shown in Figure 7.2. Clearly the operating time of the fuse should always be less, at any current level, than the period for which the circuit can withstand the condition. This seems to be self-evident but in practice there are a number of less obvious limitations and factors which must be taken into account.

A very important factor which must be recognised is that the operating times referred to above are the total clearance times, i.e. the sum of the pre-arcing and arcing times. Now, ideally, a fuselink should be capable of carrying a current just below the minimum fusing level indefinitely and also of carrying any higher current for just less than the corresponding operating time and thereafter be in its original condition. In practice, these conditions cannot be achieved because there are changes of state before operation occurs. Once arcing has commenced it is clearly impossible for an element to return to its original form and, even in the molten state, an element may distort and not return to its initial shape on cooling. Furthermore unacceptable changes may occur in fuses with low-melting-point materials on the elements, if over-currents continue long enough to initiate the M-effect diffusion process.

There is thus a deadband below the time/current characteristic in which a fuse should, if possible, not be called upon to perform. The likelihood of high current flowing through a fuselink for just less than its operating time is small but, if it should happen, the probable result, which must be accepted, is that the fuse will operate more quickly than expected on a future occasion. Difficulties would certainly arise if a fuselink carried currents near its minimum fusing level for long periods, and, to avoid this situation, a fuselink is assigned a rated current somewhat below the minimum fusing value. The ratio of the minimum fusing current to the rated value, which is defined as the fusing factor, usually has values in the range 1·25–2. The significance of this factor is that protected circuits must be able to operate continuously at levels appreciably above the rated current of the fuse if the first criterion above, namely that the circuit continuous rating must exceed the minimum fusing current, is to be satisfied. This is an uneconomic situation which arises with many protective arrangements, because it is necessary to have current settings above the full-load value of the circuit unless discrimination is achieved by differential methods. It is clear that it is particularly desirable to use fuselinks of low fusing factor when the cost of the protected circuit and equipment rises significantly with its current-carrying capacity. Another factor which must be borne in mind is that many fuselinks do not provide full-range protection, i.e. they will not operate satisfactorily at all current levels from their rated breaking capacities down to the minimum fusing values. As stated earlier, satisfactory arc extinction may not be achieved in some fuselinks at relatively low over-currents. Care must always be taken to see that such fuselinks are only used in applications where currents of these magnitudes will not be experienced or, if this cannot be guaranteed, then an associated protective scheme must be provided to interrupt these currents before the fuse can operate.

7.1.2 I^2t

It has been made clear, in Chapters 2 and 3, that fuses operate extremely rapidly at very high currents, clearance times of a few milliseconds being usual. The behaviour at such currents is obviously dependent on the current waveshape and thus factors such as the instant in the voltage cycle at which a short circuit occurs affect performance. In these circumstances, a single operating time can only be assigned to a particular RMS value of prospective current if its waveshape is defined and such values are not of great assistance in determining the suitability of a fuse for a given application because they do not take account of the current-limiting property of the fuselink. As a result, use is made of a quantity, termed I^2t, which is the time integral of the square of the instantaneous current passing through a fuselink between the instant when a circuit fault occurs and the instant of extinction of the fuse arc, i.e.

$$I^2t = \int_0^t i^2 \, dt$$

If the resistances of the items in the circuit into which the current is flowing remain constant throughout the period of operation of a fuselink, then the value of I^2t is proportional to the energy fed to the circuit. In practice, the resistances generally increase, some of them quite significantly, because of the heating produced by the over-currents. Nevertheless, the value of I^2t is described as the let-through energy, a term which is not strictly correct in any case, as it does not contain a resistance value and thus is not in energy units.

To assist in the application of fuses, many devices, such as semiconductors, are tested by the manufacturers to determine their I^2t-withstand values and these are published. Fuse manufacturers provide the I^2t values needed to operate their fuselinks at very high current levels and this makes the choice of appropriate fuselinks much simpler than if time/current curves have to be matched. Care must nevertheless be exercised when selecting fuses in this way because in practice the withstand I^2t values of components and the operating I^2t values of fuselinks both depend on current level and waveshape. They do, however, both tend to become constant at very high currents.

In practice, the manufacturers of fuses provide not only the I^2t values associated with the total clearance time, including the arcing period, but also the I^2t values let through during the pre-arcing period. Both sets of values vary with the severity of the test circuit conditions and to permit co-ordination of fuses, which is discussed in Section 7.2, to be achieved, the maximum total I^2t value and minimum pre-arcing I^2t values are quoted.

The calculation of the minimum pre-arcing I^2t for smaller ratings of a series of current ratings which have elements of the same form and material in a homogeneous series (see Glossary of Terms) can be evaluated from the following formula:

$$\left(I^2t\right)_2 = \left(I^2t\right)_1 \times \left(\frac{A_2}{A_1}\right)$$

where $\left(I^2t\right)_2$ is the pre-arcing I^2t for the smaller rating; $\left(I^2t\right)_1$ is the pre-arcing I^2t for the largest rating; A_2 is the minimum cross-sectional area of the element of smaller rating; and A_1 is the minimum cross-sectional area of the element of the largest rating.

The operating I^2t at reduced voltage than those tested can be estimated using the following formula:

$$\text{operating } I^2t \text{ at reduced voltage } V_r = \left\{ \frac{\text{Operating } I^2t \text{ at test voltage } V_t}{\text{pre-arcing } I^2t} \right\}^{V_r/V_t}$$
$$\times \text{ pre-arcing } I^2t$$

The experts on the IEC low-voltage fuse committee refer to this as the 'Henry Turner Formula' in memory of one of the pioneers in fuse technology.

7.1.3   Virtual time

As stated at the beginning of the previous section, a fuselink will have a range of operating times at each particular high value of prospective current, making it impossible to produce a single characteristic relating current and actual operating time.

To assist, a quantity termed virtual time was introduced. This is defined as the I^2t value divided by the square of the prospective current. Again two values can be produced for any prospective current level, one in which the maximum total I^2t is used and a lower one based on the minimum pre-arcing I^2t.

Virtual times are no longer used to any great extent by fusegear engineers, co-ordination being normally achieved by using I^2t values.

7.1.4   Published time/current characteristics

The data required for producing these characteristics are obtained by testing fuselinks which are at ambient temperature (15–25°C) when current flow through them is initiated. The curves published by the fuse manufacturers usually show the relationship of the pre-arcing time to prospective current.

It will be appreciated that the effects of factors such as current limiting, the instant of fault occurrence and the current waveshapes do not significantly affect

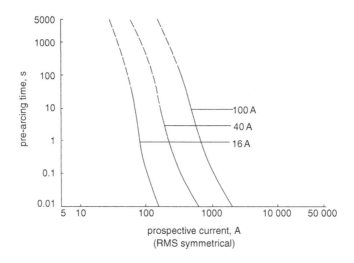

Figure 7.3   Time/current characteristics

the performance at the lower current levels where the operating times are long and, of course, the arcing periods are negligible compared with the pre-arcing times. As a consequence, the variations in operating times are insignificant at these levels and there is therefore only one time/current curve for each fuselink.

In practice, the passage of load current prior to the over-current condition or a high-temperature environment causes operation to be slightly faster than shown on the characteristic. There is also an associated reduction of the times for which the protected circuit can withstand given over-currents if it is in an environment of relatively high temperature or if load currents have been flowing in it. The two effects will not usually be equal but they nevertheless reduce the significance of this factor.

It is common to indicate the approximate minimum-breaking current on the characteristics by using a full line above this level and a broken line below it. This is illustrated in the typical characteristics shown in Figure 7.3.

Fuselinks which have minimum breaking currents above those at which their elements melt are referred to as 'back-up' or 'partial-range' fuselinks. They are mainly used to protect high-voltage cables and transformers, semiconductor devices and motor circuits.

Low-voltage industrial and domestic, high-voltage full-range and miniature fuselinks are capable of interrupting safely any fault current which will cause the fuse element to melt. Such low-voltage industrial and domestic fuselinks are given the classification 'g'. In addition, most of these fuselinks have internationally standardised characteristics and are classified by a second letter 'G' giving 'gG' as the complete designation.

Miniature fuselinks have a series of standardised characteristics which are described in Chapter 6.

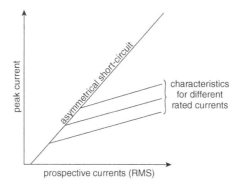

Figure 7.4   Cut-off characteristic

7.1.5   Cut-off characteristics

These show the highest possible instantaneous values of current which a given current-limiting fuselink will pass under fault conditions for varying values of prospective current. They are of use in calculating the peak mechanical forces which the equipment in protected circuits must be able to withstand.

A typical characteristic is illustrated in Figure 7.4.

7.1.6   Operating frequency

The characteristics of fuselinks normally relate to operation at frequencies of 50 or 60 Hz and there is little difference between the behaviours obtained at these frequencies. Certainly, a fuselink tested at a frequency of 50 Hz will be entirely suitable for use at 60 Hz. Even higher power frequencies would present little difficulty for the fuselink. Lower frequencies do, however, need careful consideration since extension of the duration of the half-cycles of the source-voltage wave causes the total operating times at very high currents to be increased. As a result higher arc energies are released in the fuselink. For frequencies below 50 Hz some derating in terms of rated operating voltage is therefore necessary. In the extreme case of DC applications, the voltage rating may be only half that allowable at a frequency of 50 Hz and high values of circuit inductance may necessitate further voltage derating.

7.1.7   Application of fuses to DC circuits

When a fault occurs in a DC circuit, the current rises exponentially with a time constant equal to the ratio of the inductance L to the resistance R present in the circuit, i.e.

$$i = \frac{V}{R} - I_0 \exp(-Rt/L)$$

in which I_0 is the current in the circuit at the instant of fault initiation. In the majority of circuits, the time constants are in the range 5–50 ms. As a result, the power inputs to fuse elements rise relatively slowly after the occurrence of faults, the rate of

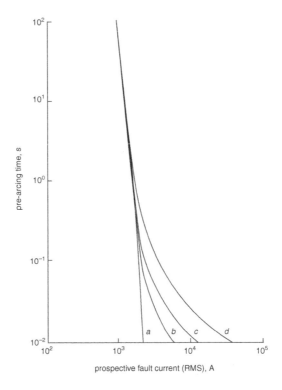

Figure 7.5   DC time/current characteristics

Where

a   symmetrical alternating fault current

b   direct-current 10 ms time constant

c   direct-current 25 ms time constant

d   direct-current 80 ms time constant

rise decreasing with increase of the circuit time constant. The times taken to cause melting of fuse elements can therefore be considerably greater than those which would be obtained if symmetrical sinusoidal currents with the same RMS values as the prospective DC values (V/R) flowed.

The above effect increases with the prospective current, i.e. when the pre-arcing times are short relative to the circuit time constants. The behaviour is illustrated in Figure 7.5 from which it can be seen that the effect becomes negligible at the lower levels of prospective current.

The longer pre-arcing times associated with high prospective direct currents in circuits with long time constants allow more energy to be dissipated from the fuse elements to the surrounding filler and therefore the I^2t inputs required to cause melting are somewhat higher than those required for the same prospective alternating currents. The I^2t inputs at particular values of direct current do not, however, rise in direct proportion to the pre-arcing times, because of the relatively slow rises of the direct currents after faults occur.

After melting of one or more of the restrictions in a fuse element has occurred, arcing commences, causing erosion of the element material and lengthening of the arc or arcs. In DC applications, however, there are no natural current zeros at which arc extinctions can occur and therefore the arcs must continue to lengthen until the voltage drops across them cause the currents to fall to very low levels at which arc extinction can occur.

As a result, the arcing durations and total operating times of fuses used in DC circuits increase with the circuit supply voltages; also the time constants of the circuits increase because the circuit inductance reduces the rate of current reduction.

Because of the above factors, manufacturers often reduce the voltage ratings of AC fuselinks which are to be used in DC circuits and they relate the voltage ratings to the circuit time constant, an example being shown in Figure 7.6.

It will be appreciated from the above that the I^2t input needed to cause operation of a fuse at a high direct current is higher than that required to interrupt an alternating current of the same RMS value.

The time constants for some typical low voltage applications are:

Application	Time constant (ms)
Battery supplies for UPS	<5
Industrial DC control and load circuits	<10
DC motors and drives	20–40
Traction catenary	up to 50
Traction, third rail	50–100
Magnets and field supplies	up to 1000

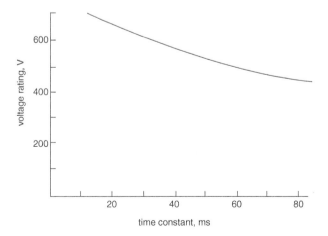

Figure 7.6   Voltage rating and time constant

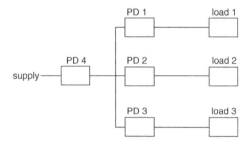

Figure 7.7   Simple network

7.2   Discrimination and co-ordination

Most circuits contain several protective devices and some of these are effectively in series. They must all be co-ordinated so that correct discrimination is achieved under all fault conditions and only the minimum of interruption should occur to clear any fault condition.

The co-ordination of the protective devices is affected by their operating characteristics and there are several possibilities which may arise; for example, a network may contain a considerable number of fuses which must be chosen to discriminate or alternatively a fuse may have to operate in series with a circuit breaker which is tripped by a protective relay. These situations are considered separately below using the simple network shown in Figure 7.7.

7.2.1   Networks protected by fuses

It is very common to employ the radial system shown in Figure 7.7 and to use a major or upstream fuse in the supply connection (PD4) and minor or downstream fuses in the individual load circuits (PD1–PD3).

Clearly, each minor fuse must have the time/current characteristic needed to protect its load circuit and a fault on a particular load should only cause its associated minor fuse to operate. The major or upstream fuse (PD4) will also carry the fault current but it must not operate or be impaired.

For faults which cause relatively small currents to flow, the arcing times, as proportions of the pre-arcing times, are small and consequently discrimination can be predicted by comparing the time/current curves of the major and minor fuses. Provided that the curves for the minor fuses are to the left of that for the major fuse, i.e. the minor fuses operate more quickly, then discrimination should be obtained. A significant margin should nevertheless be allowed for fuse tolerances and because the major fuselink may be carrying currents fed to healthy circuits as well as the fault current. If these load currents may be large, then calculations should be done to determine the possible currents in the major fuselink for given fault currents and adequate time differences should still be present between the operating times of the fuses concerned. This situation is illustrated in Figure 7.8.

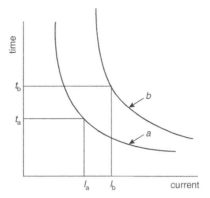

Figure 7.8 Fuselink characteristics

Where
a   minor fuselink characteristic
b   major fuselink characteristic
I_b current in major fuselink when fault current of I_a flows in minor
        fuselink
t_b must exceed t_a

At higher fault-current levels which will result in melting of the minor fuse in less than 100 ms, the arcing time of the minor fuse must be taken into account. This is done not by considering the actual values of time, but by using the I^2t values. The requirement is that the pre-arcing I^2t of the major fuse shall exceed the total operating I^2t of the minor fuse by a reasonable margin (say 40 per cent). It will be appreciated that load currents flowing in healthy circuits, while a fault exists, have negligible effects on the operation of the fuses when the fault current is very great. Prefault conditions are important, however, and the I^2t margin suggested above should be increased if it is known that, before a fault, the minor fuse is likely to be much less loaded, as a proportion of its rated current, than the major fuse. This is because the minor fuse will be operating at lower temperatures than the major fuse at the instant of fault occurrence. A suggested rough, but general, guide for the extreme case of the minor fuse being on no load and the major fuse on full load when the fault occurs, is that the rating of the major fuse should be increased by a further 25 per cent.

The particular case which arises when discrimination has to be achieved between fuses on the two sides of a transformer is discussed in Section 7.5.

7.2.2 Networks protected by fuses and devices of other types

Here the general requirement is similar to that for discrimination between two fuselinks, in that only the downstream device is required to operate. It is this latter device which has to be chosen first, because its time/current characteristic must provide the necessary protection for its associated circuit. Thereafter the upstream device must have a characteristic which will ensure discrimination.

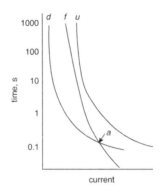

Figure 7.9   Discrimination between a current-limiting fuselink and other protective devices

Where
d   downstream-device characteristic
f   fuselink characteristic
u   upstream-device characteristic

In practice, two alternative arrangements are encountered, one in which the upstream device may be a fuse whilst the downstream devices may be small or miniature circuit breakers incorporating over-current protective features, and the other in which an upstream circuit breaker and downstream fuses are used. The characteristics associated with these two situations are shown in Figure 7.9.

With the first arrangement, there is always an actual or potential upper limit to the fault current at which discrimination can be obtained. This is because the circuit breaker or other downstream device always has a definite minimum operating time resulting from the delays in the over-current-detection equipment and the circuit breaker itself plus its own arcing time, of which the latter is not likely to be less than the duration of one half-cycle. The operating time of the upstream fuse, on the other hand, decreases continually with increase in current and the upper current limit at which discrimination can be achieved is at the intersection of curves d and f in Figure 7.9.

With the second arrangement, the curves f and u in Figure 7.9 are relevant. It will be seen that they approach each other most closely at times of the order of 1–3 s where the influence of arcing time is negligible and there is usually little difficulty in choosing characteristics which enable full discrimination to be obtained.

7.2.3   Co-ordination between a current-limiting fuse and a directly associated device of lower breaking capacity

It has already been stressed that some current-limiting fuselinks are unable to clear currents in a range above the minimum fusing level and that an associated device of limited breaking capacity is needed to interrupt these currents. In these circumstances the requirements are quite different from the preceding two cases. The current-limiting

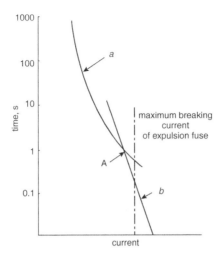

Figure 7.10   Co-ordination of a fuse and a directly associated device
Where
a   expulsion-fuse characteristic
b   current-limiting-fuse characteristic

fuse is used as a back-up for the other device. Thus in the event of heavy faults only the current-limiting fuse is required to operate while only the associated device is required to operate in the event of overloads or small faults. This is achieved by choosing characteristics for the fuselink and other device so that they produce a composite characteristic of the form shown in Figure 7.10 and this clearly must give sufficiently rapid clearance at all current levels to protect the associated circuit adequately. The other criteria which must be satisfied are as follows:

(*a*)   The take-over point (A) at which the curves intersect must be at a current level below the breaking capacity of the other device and above the minimum value which the current-limiting fuselink can interrupt, unless it is fitted with a striker which operates the other device.

(*b*)   To deal with cases where the current-limiting fuse clears the circuit, the other device must be able to carry the maximum fault current safely and, where it may have to close on to a fault, it must have a making capacity adequate for the cut-off and I^2t let-through values of the fuselink.

7.3   Protection of cables

Miniature fuselinks are not generally called upon to protect cables. High-voltage fuselinks and those for the protection of semiconductor devices usually provide only back-up protection, i.e. they only clear high currents, when used to protect items of equipment and the associated cables. Low-voltage type 'gG' fuselinks are used extensively, however, to protect cables and in this role they are required to operate over the whole range of over-current conditions. It is clearly desirable that their performance and characteristics should ensure that the cables will not be damaged because of overloading or faults in the circuits they feed and, to this end, rules for the selection and over-current protection of cables have been drawn up and included in national wiring rules or regulations. *IEC Publication 60364* deals with *Electrical Installations in Buildings*. In the UK, the Institution of Electrical Engineers (IEE) produces the *Regulations for Electrical Installations*. The 16th Edition, which was published in 1991, is based on *IEC Publication 60364* and became *BS 7671* in 1992.

The 16th Edition of the IEE Regulations includes the requirements for the over-current protection of cables and these will be met in Britain and in other countries which follow IEC practice. In these regulations, the term 'over-current' covers both short-circuit currents and overloads, an overload being defined as an over-current which flows in a circuit which is perfectly sound electrically. Clearly an overload can occur, for example, if a motor is stalled or caused to run slowly because of the torque required of it.

The first important factor which must be considered is the current-carrying capacity of the cables to be protected. This is clearly dependent on the conductor and insulation materials and dimensions. In addition, it is affected by the ambient temperature of the environment in which the cables will operate and on the installation arrangements, including the spacing and adequacy of air circulation. The current-carrying capacities of cables under a range of operating conditions have been determined and they are tabulated in the wiring regulations referred to above.

To avoid damage it is essential that the maximum sustained current (I_B) carried by a cable should be less than or equal to its current-carrying capacity (I_Z), i.e.

$$I_B \leq I_Z$$

To allow the maximum sustained current to flow, the fuse must have an equal or higher rated current I_n and to provide adequate protection the fuse rating should not exceed the current-carrying capacity of the cable; therefore:

$$I_B \leq I_n \leq I_Z$$

A cable can carry currents above its current-carrying capacity I_Z for limited periods and to give ideal protection, the protecting fuselinks should operate, at any current level, within the period for which the cable can carry that current. Strictly, to eliminate all possibility of damage, the minimum fusing current of a fuse should be just below the current-carrying capacity of the protected cables so that the time/current characteristics will nest as shown in Figure 7.11. This would require, for example, the installation of cables able to continuously carry currents up to 50 per cent more than the fuse rating, if the fuses had a fusing factor of 1·5 and, of course, as stated above, the maximum-sustained current would not have to exceed the fuse rating.

Such an arrangement could be unacceptably costly and, in practice, slight risks are taken. The limiting factor is the temperature reached by the insulation as a result of conductor heating during overloads of fairly long duration, of the order of 1 h or more, depending on the thermal-time constant of the cable. In general this period

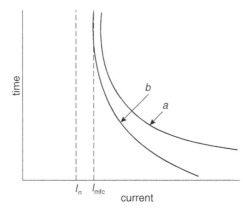

Figure 7.11   *Cable and protective fuse characteristics*
Where
a   cable-withstand characteristic
b   fuse-operating characteristic
I_{mfc} minimum fusing current of fuselink
I_n rated current of fuselink

increases with cable size and current-carrying capacity. The regulations, which are intended to ensure that the life of the insulation is not significantly shortened, specify that the minimum-operating current of the protective devices should be equal to or less than 1·45 times the current-carrying capacity of the cable (i.e. $1{\cdot}45I_Z$).

In order to verify that 'gG' fuselinks are capable of protecting cables against overload, a conventional cable-overload-protection test has been introduced into *IEC 60269-1*. The fuselink must operate when subjected to an overload current of 1·45 times the rated current of the cable it is to protect ($1{\cdot}45I_Z$), the value of I_Z being taken from a common installation condition. In practice, several fuselinks will be fitted into an enclosure so that the 1·45 I_Z condition will be met even when the fuse and cable ratings are equal ($I_n = I_Z$).

Semi-enclosed fuselinks require about twice their rated current to give operation in periods of one hour or more and therefore to meet the 1·45 condition:

$$I_n \leq \frac{1{\cdot}45}{2} I_Z$$

This means that larger cables may have to be used if semi-enclosed fuselinks are used for cable protection rather than cartridge fuselinks.

The wiring regulations are not framed to allow for frequent overloads because such over-currents could shorten the life of the cable dielectric. Advantage is not therefore to be taken of the overload capacity of cables and, if frequent overloads are likely to occur, then these currents should be regarded as the normal current of the circuit and correspondingly larger cables should be installed.

When the fuselinks are selected on the above $1\cdot45I_Z$ basis, the shape of the time/current characteristics ensures that the cables are adequately protected at higher over-currents.

In those applications where the low-voltage fuselinks are to provide back-up or short-circuit protection to the cables, then co-ordination must be ensured by providing fuselinks which let through I^2t values lower than those which can be withstood by the cables. For fault durations of 5 s or less the I^2t withstand of cables may be determined from the expression

$$I^2t = K^2a^2$$

in which a is the cross-sectional area of the cable conductor in square millimetres and K is a factor which depends on the conductor material and the limiting temperature which can be withstood by the insulation. Values of K for various conductor and insulator combinations are given in the current edition of the IEE Regulations. The values range from 76 for aluminium conductors insulated with PVC material to 143 for copper conductors with 90°C thermosetting insulation.

Co-ordination is normally checked using the fuselink I^2t value associated with operation in 5 s.

It will be noted that the I^2t withstand of the cable is not affected by the duration of the short circuit. That of the fuselink does increase with operating time, however, and therefore correct operation can be assured by checking that the fuselink I^2t value associated with interruption in 5 s is lower than the cable-withstand value.

The requirements for the selection of fuses for the protection of conductors are found in the North American wiring regulations and are summarised below and taken from *IEC TR 61818*.

(a)   The voltage rating of the fuse is selected to be equal to or greater than the maximum system voltage.

(b)   The load current is calculated and multiplied by $1\cdot25$ for continuous loads (continuous loads are those which are present for 3 h or more).

(c)   The conductor size is selected from an ampacity (current-carrying capacity) table found in the wiring regulations.

(d)   The general rule for selecting the fuse is to select a standard fuse current rating to coincide with the conductor ampacity. For conductor ampacity less than 800 A, if the conductor ampacity falls between two standard fuselink current ratings, the larger fuselink current rating is used. For conductor ampacities of 800 A and over, if the ampacity falls in between two standard fuselink current ratings, then the smaller fuselink current rating is used.

(e)   The fuse is selected to protect the conductor under short-circuit conditions. In practice, North American cable standards have been co-ordinated with fuse standards so that short-circuit protection is achieved. For other types of conductors, short-circuit withstand ratings are compared with the fuse characteristics to make sure that conductor damage does not occur.

7.4   Protection of motors

High-voltage and low-voltage current-limiting cartridge fuselinks are used in conjunction with either air-break or vacuum contactors in many countries to protect three-phase AC induction motors up to 2 MW rating operating at line voltages up to 11 kV. The fuses, as stated earlier, provide the protection against short circuits and must therefore have adequate capacities. The lower currents are cleared by the overload protection in the motor starters. In these circumstances the rated current of the fuselinks does not need to correspond to the motor rating, and certainly when motors with direct-on-line starting are to be protected the choice of fuse-current rating is dictated by its ability to withstand the motor starting-current surge, typically 5–6 times the full-load current. This usually results in the use of fuselinks with rated currents up to twice the motor full load current.

Such fuses thus carry up to about three times their rated current during starting periods. Consequently, and because of the low thermal inertia of the fuse elements, they reach temperatures considerably higher than those caused by continuous operation at their rated current. The resulting expansion and contraction would tend to lead to mechanical failures in long fragile elements, after a number of motor starts, and therefore the elements of high-voltage motor fuselinks are corrugated, as stated earlier, to minimise this effect and avoid the necessity of choosing fuses with even higher ratings. The elements of low-voltage fuselinks tend to be more robust and corrugation has not been found to be necessary.

For high-voltage applications *IEC 60644* gives additional withstand tests for motor circuit applications. This pulse withstand test introduces a '*K*' factor. The '*K*' factor defines the overload characteristic to which the fuselink may be repeatedly subjected under specified motor starting conditions and other specified motor operating overloads without deterioration. The overload characteristic is obtained by multiplying the current on the pre-arcing characteristic by '*K*' (less than unity). Two sequences of withstand tests are specified and both are based on the pre-arcing current at 10 s.

The surges are not so great when other methods of starting, e.g. assisted starting, are employed, and therefore fuses with lower current ratings may be used. However their temperature rises under running conditions must be considered and in general the current rating of high-voltage fuselinks should be at least 125 per cent of the rated current of the motor. Such over-rating is not usually necessary when low-voltage fuselinks are used. Allowance may also have to be made for the high transient currents which flow, with some methods of starting, when transitions are made from one connection to a succeeding one, as occurs, for example, when a star–delta starter or a rotor–resistor starter is used.

Because motor fuselinks do not operate continuously at their rated currents, it is often possible and is common practice to mount low-voltage fuselinks of given ratings in carriers and bases of lower current ratings and this of course limits the maximum continuous current capability of the combination to the lower level. As an example, a fuselink with operational characteristics of 50 A may be mounted in a fuse holder with a rating of 32 A. *IEC 60269-2-1 Sections II and IV (BS88-2 and -6)* recommends that this combination be designated 32M50, the first figure indicating the maximum continuous current rating of the complete fuse. This type of fuselink has been given the classification 'gM'.

Continental European practice also has a special type of low-voltage back-up fuselink for use in motor circuits and it is designated 'aM', see Section 8.1.8.1. An 'aM' fuselink is selected with the next higher fuse rating than the full load current of the motor, thus making fuse selection more straight forward.

In North America, there are two basic types of time current characteristics, 'gN' (normal) and 'gD' (time delay) which have been standardised in *IEC 60269-2-1 Section V.* The 'gN' characteristics are similar to 'gG' and the 'gD' characteristics have to withstand five times full load current for 10 s.

Examples of 'gG', 'gM', 'aM', 'gN' and 'gD' fuses for a typical direct on-line motor starting duty for a motor full load current of 28 A are shown in the following table:

Fuse type	Origin	Suitable rating
gG	General purpose IEC fuse	63 A
gM	Motor circuit fuse	32 M 63
aM	Back-up fuse	32 A
gN	North American 'normal' fuse	70 A
gD	North American 'time delay' fuse	40 A

These examples are merely illustrative and assume that the starting time is less than 10 s, that the maximum starting current does not exceed seven times full load current, and that starts are infrequent.

Correlation between the protective items associated with a motor is illustrated by the characteristics shown in Figure 7.12. This represents a typical application involving a motor, a relay or relays (providing one or more of the following: inverse over-current protection, instantaneous over-current protection, instantaneous earth-fault protection), a contactor or other switching device, the cable to the motor and the fuse itself. The motor will have been chosen for its particular duty, thus fixing the values of the full-load and starting currents. In addition, the duration and frequency of the starts will also have been fixed. With these data available, the characteristics of the associated inverse over-current relay (*b*) are then chosen to give adequate thermal protection to the motor. The switching device must then be selected in conjunction with the fuselink to co-ordinate with the motor. In particular:

(*i*)   The time/current characteristic of the fuselink (*c*) must lie to the right of the point s on the motor characteristic by an adequate margin.

(*ii*)   The current corresponding to the intersection point (H) of the fuselink and relay characteristics (*c* and *b*) must not exceed the breaking capacity of the switching device.

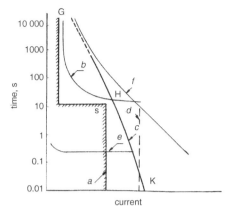

Figure 7.12 *Characteristics associated with the protection of a motor*
Where
a motor starting current
b overload-relay characteristic
c fuselink characteristic
d breaking capacity of switching device
e instantaneous relay characteristic
f cable-withstand characteristic

(*iii*) In the event of an instantaneous relay being fitted, with the characteristic *e*, the switching device might be called upon to clear currents up to the level of the intersection point (J) of the characteristics *c* and *e*.

(*iv*) As stated in Section 7.2.3, the cut-off current of the fuselink should not exceed the through-fault-current withstand level of the switching device for the operating time of the fuselink.

(*v*) The minimum breaking current of the fuselink should be low. It must be less than the current co-ordinate of point H. When possible it should be at least as low as the starting current of the motor, thus ensuring that, in the event of a relay failure, the fuselinks can offer protection against a locked-rotor condition.

(*vi*) The whole of the withstand curve of the cable (*f*) must lie to the right of the operating characteristic GHJK as shown in Figure 7.12. Where fuselinks with high rated currents are needed because of the nature of the starting duty (for example, long starting times and/or frequent starts) the section HJK of the operating characteristic provided by the fuselink moves to the right and this may necessitate the use of a cable of larger cross-sectional area.

For low-voltage applications the requirements for contactors and motor starters are given in *IEC 60947-4-1* which was first published in 1990. This includes the co-ordination requirements with short-circuit-protective devices (SCPDs). The rated conditional short-circuit currents of the contactors and starters backed up by the SCPDs are specified. Two types of co-ordination are permissible: type '1' and type '2'.

Type '1' co-ordination requires that, under short-circuit conditions, the contactor or starter shall cause no danger to persons or installation and may not be suitable for further service without repair and replacement of parts.

Type '2' co-ordination requires that, under short-circuit conditions, the contactor or starter shall cause no danger to persons or installation and shall be suitable for further use. The risk of contact welding is recognised, in which case the manufacturer shall indicate the measures to be taken as regards the maintenance of the equipment.

Clearly type '2' is the preferred co-ordination. In the past decade there have been developments in contactors and motor starters which have required the short-circuit-protection device to have relatively low values of let-through I^2t and cut-off characteristics. Co-ordination recommendations are made by the manufacturers of motor starters in accordance with *IEC 60947-4-1*.

The low-voltage fuse committee of the IEC produced the *Technical Report 61459* in 1996 – Application Guide for Co-ordination between fuses and contactors/motor starters. It reports that from a survey of tests, that type '2' co-ordination is achieved by using 'gG' or 'gM' fuselink which have pre-arcing I^2t values towards the bottom of the limits specified in *IEC 60269-1*.

7.4.1 Protection of soft starters

AC induction motors are frequently used at fixed speeds and traditionally these have been started at full rated voltage, direct on line (DOL). This is still very popular up to 7.5 kW. However, DOL starting gives far higher torque than is delivered at full speed and creates a 'jolt' in the motor that can result in wear and mechanical damage to the motor, gearboxes, clutches, couplings, transmission equipment and the load including goods being handled. The high associated start-up currents can also cause significant line voltage dips affecting the power quality of the system, see Section 7.11.

As indicated earlier, there are other methods of assisted starting, including star–delta, auto-transformer and pole changing motors. With the advent of economical and reliable power semiconductors, there has been an ever-increasing use of 'electronic soft starters'. The modern soft starter usually consists of six thyristors arranged in anti-phase parallel configuration, as shown in Figure 7.13. This is the most common connection method, however for large motors, up to 1000 kW the thyristors are sometimes connected in the delta circuit thus reducing the thyristor current to 58 per cent.

When the soft starter is activated, the thyristors will switch out large parts of the supply voltage, gradually less and less of the supply voltage is switched out until the full voltage is supplied to the motor. Slow ramping up the voltage avoids both current surges and torque transients. The soft starter has adjustments for limiting the starting current and setting the ramp up time. Typically starting current is reduced to 300 per cent of the full load current with ramp up times approaching 30 s.

Soft starters can also cover torque control starting, low speed jogging, kick starting, soft stopping and breaking. In addition, they can provide remote communication.

In some designs, the thyristors are by-passed via a contactor when the motor is run up to speed, so that the mains voltage is applied directly to the motor and the thyristors are not in continuous operation reducing size and increasing efficiency.

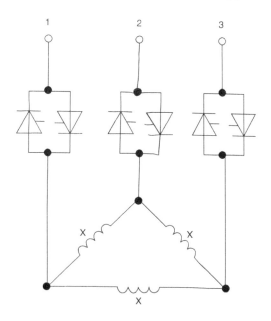

Figure 7.13   Typical soft starter connection method

Like their electromechanical counterparts electronic soft starters need to be protected against high over-currents with an SCPD. It will be seen in Section 7.8 that power semiconductors can be damaged by high over-currents and that special fast-acting fuses are required. *IEC 60947-4-2* covers type 1 and type 2 co-ordination in a similar way to that described earlier for electromechanical contactors and motor starters to *IEC 60947-4-1*.

The fuselinks are normally fitted in the input supply lines. The current rating of the fuselink selected is often governed by the repetitive duty of the motor, taking into account the cyclic loading factor for the fuselink, as described in Section 7.8.4. The I^2t let-through by the fuselink then needs to be less than the I^2t withstand of the power semiconductor to give type 2 co-ordination.

7.5   Protection of power transformers

Both step-up and step-down power transformers are used in power systems but almost all of the former type are used in conjunction with the alternators in generating stations to form generator–transformer units or they are used to interconnect main transmission networks. In both these applications, differential-type relay protective schemes are employed because of the VA levels involved. Step-down transformers are much more widely used and there are many in the distribution networks which rely on fuse protection on both the high-voltage (supply) side and low-voltage (load) side.

In the UK, high-voltage fuses are connected in the primary circuits of three-phase, 11 kV/415 V distribution transformers rated up to 1 MV A and low-voltage fuses are connected in the secondary circuits. Transformers operating at the same voltages are used in industrial premises, but these may be somewhat larger, ratings up to 1·5 MV A being quite common, and these are also protected by fuses in the high- and low-voltage circuits.

It is common for the above transformers to be fed by cable, in which event current-limiting cartridge fuses positioned near the transformers are used. In some cases, however, the supply is fed by an overhead line to one, or even a group of, transformers, and then the latter and the line are protected by a set of fuses at the input end of the line. High-voltage fuses are also included on both sides of three-phase, 33/11 kV transformers with ratings up to 5 MV A.

The normal purpose of the fuses on the lower-voltage side of a transformer is to protect the load circuits connected to the secondary windings and in these circumstances the characteristics of the fuselinks have to be chosen suitably to match those of the loads and connecting cables. If, however, there is interconnection in the lower-voltage network because of the parallel connection of transformers, then the secondary-circuit fuses may carry currents being fed back into a transformer in the event of a fault within it, and fuses capable of operating under such conditions, as well as providing protection for the loads, must be chosen.

The fuses on the higher-voltage side of a transformer must isolate it if a fault occurs within it, and this must be done with the minimum disturbance to the system and without causing unnecessary loss of supply to the healthy parts of the system. When choosing these fuse-links, the following factors must be considered:

(*a*) It is common in some countries, including the UK, deliberately to operate power transformers above their continuous ratings, for predetermined periods, which may be of several hours, this being possible because of their relatively long thermal time constants. An individual phase may be further overloaded at any time owing to unbalanced loading, the limiting factor being the total power loss dissipated by all the windings and the core. To avoid incorrect operation of the fuses it is usual to use ratings capable of continuously carrying the maximum currents that may flow under the above conditions and, in doing this, account must be taken of any derating which may be necessary when the fuselinks are mounted in small enclosures, to ensure that the permissible temperature rise limits of both the fuselinks and enclosures are not exceeded. Derating may also be needed when the fuselinks are to experience high ambient temperatures.

(*b*) Transformers, like motors, may draw very large transient currents when they are energised. This is because their cores may be driven deeply into saturation for part of each cycle of the power frequency and this is accompanied by a magnetising-current inrush which has a unidirectional component and a waveshape of the form shown in Figure 7.14. Although the total exciting current of a transformer is only 2 or 3 per cent of the rated current in the steady state, the initial transient values may be many times the rated current and their time constants may be quite long. The actual values depend on the transformer design, the maximum system voltage that may be

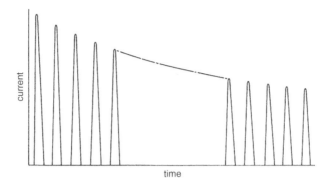

Figure 7.14   Magnetising current of power transformer

present and on the fault current available at the transformer, this being related to the circuit impedance. The fuselinks should not be operated by these surges and strictly the values to be withstood should be ascertained. In practice, the value of inrush current, as a proportion of transformer-rated current, tends to fall with increase of the kVA rating. On the other hand, the duration of the inrush transient tends to increase with kVA rating. Since these factors vary oppositely, it is possible with distribution transformers, when actual values are not available, to assume that the inrush current is equivalent to either 10 or 12 times the transformer full-load current for a duration of 100 ms. This rule has been in general use for some time. It is sometimes supplemented by the requirement that melting of the fuselink elements should not start in less than 10 ms when carrying 25 times the transformer-rated current. The higher multiple (12 times) is used for fuselinks without strikers to provide sufficient margin to ensure that the fuselinks will not melt towards the end of the magnetising-current surge. Such an occurrence would be unacceptable because the fuselinks might not be able to withstand the system voltage and would then continue to arc while carrying the lower value of follow-on current. Failure or interruption would eventually occur after several minutes or hours.

When fuselinks with strikers which are arranged to trip associated switches are being considered, it may then be permissible to accept the risk of a very occasional spurious fuse operation and so the lower multiple of ten times may be acceptable. Clearly if melting does occur towards the end of a surge, switch operation will be initiated and clearance will be effected. In this connection it will be realised that the highest value of magnetising-current inrush seldom occurs because it is associated with energisation at particular instants in the voltage cycle.

It is clear that the time/current characteristic of the high-voltage fuselinks should pass to the right of the above points, as shown in Figure 7.15, to ensure non-operation.

(*c*) On systems which are electrically exposed, over-voltages resulting from lightning can be impressed on transformers and, even if they do not cause any damage, high currents will nevertheless flow into them. Ideally the fuses should not operate in these circumstances, but to achieve this they may require such high ratings that

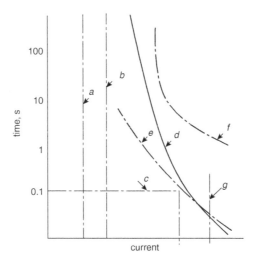

Figure 7.15   Characteristics associated with the protection of a transformer
Where
a   full-load current of transformer
b   permissible overload current of transformer
c   magnetising-inrush equivalent current
d   HV fuselink characteristic
e   LV fuselink characteristic (referred to HV side)
f   characteristic of source circuit-breaker relay
g   maximum current on HV side with fault on LV side

the cover at other times would be inadequate. Usually some risk has to be accepted and practice varies widely, decisions being taken in the light of the experience of the particular users involved.

(*d*) To provide rapid clearance of faults within a protected transformer, the current required for operation in the 10 s region of the fuselink time/current characteristic should be as low as possible. This also enables good co-ordination to be achieved with the over-current protective devices in the high-voltage supply network.

(*e*) The minimum breaking current of the high-voltage fuselinks should be as low as possible to enable them to clear most of the faults which may occur within the transformer. There are inevitably some conditions, such as inter-turn faults, which may cause primary currents of less than the maximum full-load level to flow and these, of course, cannot be detected or cleared by the fuses.

Faults involving the short circuit of one or more turns may cause primary currents just greater than the minimum-fusing level of the fuselinks and unless full-range fuses are used, satisfactory interruption might not be achieved unless the fuses have strikers which will initiate the opening of an associated switch.

Another situation, which can lead to the same difficulty, can arise if a single-phase earth fault occurs on an otherwise unearthed section of a network, as illustrated in

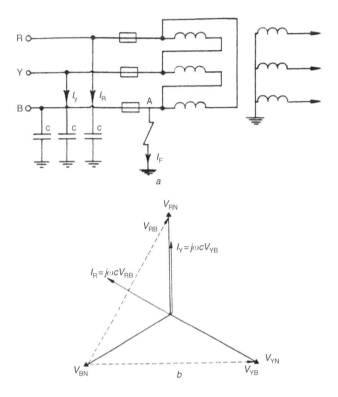

Figure 7.16   Earth fault on an insulated network

Figure 7.16a, because the fault current must return through the capacitance to earth of the network and may thus be very limited. For a fault at point A, the capacitance of phase B is short-circuited and the fuselink in phase B carries the load current and the sum of the capacitance currents then flowing to ground from phases R *and* Y. These total to three times the normal capacitive current per phase as can be seen from Figure 7.16b. This fault current is likely to remain steady and may be of a value which will melt the fuselink element after an appreciable time. Full-range fuselinks are again desirable unless striker-initiated tripping is provided.

(*f*) Correct discrimination with protective equipment and other fuses in the network, including those on the low-voltage side of the transformer, should be achieved under all conditions. Account must be taken of the fact that the transformation ratio is not fixed if tap-changing facilities are provided and, in addition, allowance must be made for the different current distributions that can exist in the primary- and secondary-phase windings when unbalanced faults are present, if the winding connections of the two sides are different, for example, star–delta. This situation is illustrated in Figure 7.17 from which it can be seen that the ratio, under normal conditions, of the currents in the low-voltage fuselinks to those in the high-voltage fuselinks is the same as the step-down ratio of the line voltages. For a fault between two of the supply

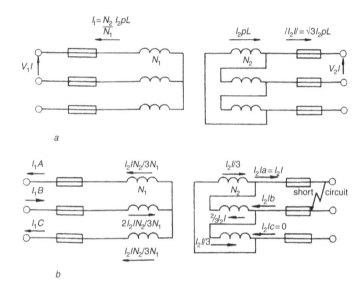

Figure 7.17   Phase–phase fault on star/delta-connected transformer

Where

a Normal condition on balanced three-phase fault

$$\text{Step-down ratio} = \frac{|V_1 l|}{|V_2 l|} = \frac{\sqrt{3}N_1}{N_2} = \frac{|I_2 l|}{|I_1 l|}$$

b Line-to-line short circuit on LV side

$$\text{Current ratios} \quad \frac{I_2 la}{I_1 A} = \frac{3N_1}{N_2} : \frac{I_2 lb}{I_1 B} = \frac{3N_1}{2N_2} : \frac{I_2 lc}{I_1 c} = 0$$

lines on the secondary side, however, the three ratios are all different and the greatest primary current has a magnitude of $2N_2/3N_1$ times the secondary line current which is also the fault current. This is 15 per cent higher than the primary current for a balanced three-phase fault of the same magnitude.

Having considered the above factors, account must be taken of the differences in the characteristics of the fuselinks on the high-voltage and low-voltage sides which arise because their basic designs are somewhat dissimilar. It is found when considering discrimination between two fuses of the same voltage rating and similar design, but with different current ratings, that the two time/current curves diverge with increasing prospective current. The contrary is the case, however, with fuses on the two sides of a step-down transformer, because the primary fuses have a much lower current rating than the secondary fuses. As a result, the time/current curve of the primary-circuit fuselinks is steeper at high currents than that of the secondary-circuit fuselinks and the curves converge with increasing prospective current as shown in Figure 7.18 and it is even possible that the curves might intersect. The attainment of discrimination is thus more difficult to achieve than with two fuselinks which are directly in series. Clearly, the intersection of the characteristics of the primary- and secondary-circuit fuselinks should not occur at currents less than those which will be present for the maximum fault level on the load circuit. To achieve this might necessitate using primary-circuit fuselinks of unacceptably high current ratings and sometimes, in the

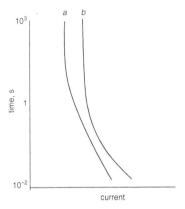

Figure 7.18   *Fuselink characteristics*
               Where
               a   LV fuse characteristic
               b   HV fuse characteristic

interests of reaching the best compromise, the risk of maldiscrimination between the primary- and secondary-circuit fuselinks for faults, on the load circuit, but very near the transformer, is accepted.

It should be noted that 'gU' low-voltage fuselinks which comply with *IEC 60269-2-1 section VI (BS 88-5)* have standardised time/current characteristics which are particularly suited to give discrimination with the standardised characteristics of the high-voltage fuselinks. To give adequate discrimination, a lower current rating would be required for 'gG' low-voltage fuselinks. This is one of the many advantages of using 'gU' low-voltage fuselinks for this application.

(*g*) When fuselinks are fitted with strikers which operate associated switches it is a requirement that the minimum breaking current of the fuselinks is less than the maximum breaking current of the switching device. In these situations the minimum breaking current is of little importance if, as is usually the case, the tripping of the switch by the striker is instantaneous. This can be understood by considering the so-called 'survival time' of the fuselink which is taken to be the time for which it can sustain, without external damage, arcing produced by a current which it cannot break. The survival time has a minimum value when the fault current is only slightly less than the minimum breaking current. In practice, the survival times are greater by a large margin, than the times from ejection of the striker to separation of the switch contacts. Typical values are:

Tripping time of switch fuse	0·04 s
Survival time	0·6 s

As a result fuselinks with quite high minimum breaking current levels may be used.

It can be seen from the foregoing that care must be exercised in selecting fuses for use with power transformers. Nevertheless, the general requirements for satisfactory co-ordination are illustrated in Figure 7.15, which shows the various time/current characteristics which must be considered.

7.6   Protection of voltage transformers

Electromagnetic voltage transformers are used to provide secondary outputs up to a few hundred volt-amperes to measuring equipment and relays at low voltages. These are standardised at 63·5 V per phase (110 V line) in the UK and most other countries.

Because the transformers are relatively small physically and thus have cores of small cross-sectional area, they have primary windings consisting of very large numbers of turns of fine wire. The step-down ratios are high and in consequence the primary windings carry currents of only a few milliamperes under normal conditions.

As stated in Section 5.1, it would be desirable to protect against all possible faults within a transformer, including inter-turn breakdowns, but this would necessitate the use of fuselinks, in the high-voltage side, with such low rated currents that their elements would be very fragile and liable to break because of mechanical vibration or shock. Such failures would lead to the de-energisation of vital protective relays and other equipment; an unacceptable occurrence. In addition, such fuselinks would not be able to withstand the excitation-current surges associated with energisation of the transformers. For these reasons, it is the practice to use fuselinks with minimum current ratings of 2–3 A for all voltage ratings. Even at these ratings, the elements are of small cross-sectional area and susceptible to the effects of corona discharge. This is especially so for ratings of 12 kV (line) and above. The effect of such discharge is to erode the elements over periods of months or years causing eventual unwanted fuse operation. It is necessary to ensure that the fuse barrels are kept well away from earthed metalwork and especially flanges with sharp edges, to reduce the likelihood of such failures.

When fuselinks are fitted on the high-voltage sides of voltage transformers, a not-invariable practice because some users prefer to rely on the main system protection, the fuselinks only operate for relatively large faults and thus disconnect the faulty transformer. They do not protect it from damage and it is expected that small faults will grow until the operating current of the fuses is reached.

Low-voltage fuses are included in the secondary circuits of voltage transformers to provide cover for faults in the burdens. There is little difficulty in choosing suitably rated fuselinks because the currents under normal conditions are of the order of a few amperes. Grading problems do not arise because of the low referred rated-current values of these fuselinks compared with those on the high-voltage side.

7.7   Protection of capacitors

For low-voltage power-factor-correction applications it is usual to install a single capacitor in each phase. Simple recommendations, associated with the protection of these capacitors, are usually provided by fuse manufacturers. These are based on

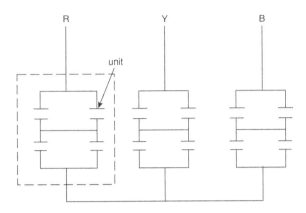

Figure 7.19   Capacitor bank

service experience and take into account the high transient inrush currents, the possible harmonic content of the currents and the capacitor tolerances. As an example, when fuselinks to *IEC 60269-2-1 Section II (BS 88-2)* are to be used to protect capacitors of ratings in excess of 25 kVA it is usual to use fuselinks with current ratings of at least 1·5 times the capacitor full-load current.

In large installations at higher voltages, capacitor banks are made up of individual capacitors connected to form a number of separate units. For three-phase applications up to 11 kV (line) and 1 MVAr, the phases may be star- or delta-connected and the units are connected in parallel. For higher-voltage systems the phases are star-connected and the units in each are connected in parallel, as shown in Figure 7.19.

A practice widely adopted in European countries apart from the UK is to fuse each individual capacitor element in the units. The fuses used for this purpose contain simple wire elements with the appropriate low-current rating and breaking capacity. The alternative practice, which is used in the UK, is to fuse each unit as a whole, although frequently a line fuse for each phase is also included for the smaller banks used at voltages up to 11 kV. A unit fuse should operate if its associated unit becomes faulty, leaving the remainder of the bank in service.

As with other applications, the requirements are that a fuse should operate as quickly as possible in the event of a fault but also be able to carry load current and transient overcurrents. The latter arise in the event of a sudden change in the voltage across a bank, a situation which arises on connection to the supply or in the event of a system fault which affects the network voltages. To prevent operation under this condition, it is usually necessary to use fuses with a current rating considerably higher than the normal capacitor current.

If a unit develops a short circuit, a discharge will occur within it and current will flow into the unit from the supply and from other healthy units, as shown in Figure 7.20. Clearly only the fuselink associated with the faulty unit should operate in these circumstances. Considering the situation with four units per phase connected in series–parallel as shown in Figure 7.21, it will be seen that the discharge current of

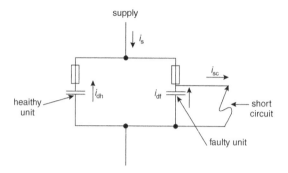

Figure 7.20   Faulty capacitor unit

Where

i_{sc} current in short circuit

i_s current from supply

i_{df} discharge current of faulty capacitor

i_{dh} discharge current of healthy capacitor

i_{sc} $i_s + i_{df} + i_{dh}$

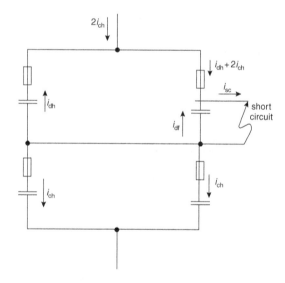

Figure 7.21   Superimposed current flow due to short circuit of a capacitor unit

Where

i_{sc} short-circuit current

i_{ch} charging current of healthy units

i_{dh} discharge current of healthy unit

i_{df} discharge current of faulty unit

i_{sc} $2i_{ch} + i_{dh} + i_{df}$

the faulty capacitor does not flow through its fuselink. Each of the two lower fuselinks carries the current needed to double the instantaneous voltage on its associated unit and the fuselink associated with the healthy upper unit carries the current needed to reduce the charge on its capacitors to zero. These three currents are all equal and should not operate the fuselinks because, as explained above, they must be chosen to withstand similar currents when the capacitors are connected to the supply. The fuselink associated with the faulty unit carries the sum of these three currents, as can be seen from Figure 7.21, and it should operate to give correct discrimination. After the initial surge and until the fuselink melts, it will carry four times the normal steady-state current.

A factor which must be considered is that the voltage on the healthy upper unit will rise to 133 per cent of its normal value and, of course, its VAR input will rise by 77 per cent.

Actual arrangements using more series- and parallel-connected units can be considered in a similar manner.

The special factors which must be considered when choosing fuses which are to protect capacitors are summarised below:

(*a*) They must not deteriorate or be damaged by the high and rapidly changing inrush currents which may flow when capacitor banks are energised or when healthy units discharge into a fault. To achieve this it is usually necessary to use fuselinks with a current rating considerably higher than the current they normally carry. In this connection it must be recognised that fuselinks of small current rating are more sensitive to inrush currents of given multiples of the rated values than larger fuselinks. Thus, the ratio of permissible fuse rating to load current decreases with increase of load current. Figure 7.22 gives a guide for the selection, based on ability to withstand inrush currents, of typical 10–300 A high-voltage fuselinks, used in air, for different values of full-load capacity current.

(*b*) The fuselinks must also be of sufficiently high current rating to withstand not only the continuous maximum load current but also the harmonic content, which can be quite significant because of the lower reactance presented to harmonics by capacitors. In practice there is a maximum permitted harmonic content and this dictates a fuse-current rating of not less than 143 per cent of the nominal full-load current of the circuit.

(*c*) Where the fuses are to be mounted in enclosures having restricted ventilation and/or where the ambient temperatures may exceed 40°C, it may be necessary to derate the fuselinks.

(*d*) Where an installation comprises banks which are in close proximity and where they are switched separately, allowance must be made for the transient inrush current which may flow between banks when one bank is to be switched in parallel with already energised banks. In practice it has been found that it is sufficient to choose the bank fuselinks by assuming the capacitor current in Figure 7.22 to be 1·6 times the actual value.

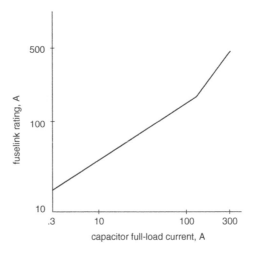

Figure 7.22   Capacitor fuselink selection curve

One or more of factors (*a*)–(*d*) dictates the smallest current rating of fuselinks which may be used in a given application, but consideration must then be given to the following factors to ascertain the degree of protection which will be obtained:

(*e*)   In situations where the unit or line fuselinks are to be used without any associated over-current-protective device, consideration should be given to the minimum breaking capacity of the fuses. This should be low enough to prevent maloperation in the event of an over-current arising owing to the short circuiting of a capacitor unit.

(*f*)   Calculations of the type described earlier should be done to determine the surge and steady-state currents which will flow when faults occur within individual units of a capacitor bank to ensure that the appropriate unit fuselink will operate and correct discrimination will be achieved. In practice it may happen that the fuselink will be operated by the high surge current but this cannot always be relied on. Consequently, the rating should be such that the steady-state power-frequency current will cause operation. In this connection it must be recognised that the fuse current will be of low leading power factor if a fault occurs in one unit of a bank which has several units in series or it will be of low lagging power factor in banks containing only parallel-connected units, the supply-system impedance then being responsible for the current limitation in the event of a short circuit. Clearance for this particular fault condition should be sufficiently rapid to limit the energy input to the container of the faulty capacitor to a level below that which could cause the case to rupture.

(*g*)   For capacitor installations with many units in parallel, the energy which may be fed to a faulty unit, by the units in parallel with it, should be calculated. The maximum energy, which may be stored by, and which must therefore be dissipated by, the capacitors in parallel with a faulty unit is given by

$$\text{maximum energy} = \tfrac{1}{2}C_{p}V_{pk}^{2} = C_{p}V^{2}$$

in which C_p is the total capacitance of the parallel capacitors and V_{pk} and V are the peak and RMS steady-state voltages across them.

The above equation is often expressed in the form:

$$\text{maximum energy} = 3 \cdot 18 \times (\text{kVAr of parallel units}) \text{ for } 50\,\text{Hz}$$
$$\text{or} \qquad\qquad 2 \cdot 65 \times (\text{kVAr of parallel units}) \text{ for } 60\,\text{Hz}$$

There is a limit to the discharge energy with which fuselinks can cope, a typical value being 10 kJ.

(*h*) As stated earlier, line fuses are sometimes included, in addition to unit and/or individual capacitor fuses, to isolate a complete capacitor bank in the event of terminal faults against which the other fuses provide no protection. Different considerations apply to these fuses and in general there is no need to use fuselinks of as low a current rating as possible because the other fuses will operate for faults within the capacitors and prevent rupture of the containers. It is the normal practice to use line fuses with current ratings at least 2·5 times that of the unit fuses to ensure correct discrimination and, of course, even higher values may be necessary if there are many units in parallel in a bank.

Both expulsion and current-limiting cartridge high-voltage fuselinks are used for capacitor protection, the former being limited to outdoor applications and where, because the banks contain units connected in series, high inductive breaking capacity is not needed. Cartridge fuses immersed in oil or air are used for all other applications.

7.8   Protection of semiconductor devices

Semiconductor power diodes were first marketed in 1953. It was realised from the outset that these devices had very limited overload capacities and, as they were expensive, the fuse manufacturers attempted to produce fuses which were more sensitive to overloads and which would operate more quickly than their conventional designs. As a result, the first applications were filed in 1955 for patents on fuses specifically designed to protect semiconductor rectifiers (e.g. US Patent 2 921 250, 13 June 1955). The invention of the thyristor and the subsequent rapid expansion of the power-electronics industry which it initiated, made the need for semiconductor fuses even more apparent.

Today, semiconductor devices are being manufactured with maximum continuous current ratings up to 15 kA and peak inverse voltages of 7 kV. Unfortunately, the devices still have poor overload capacities and continue to need sensitive and fast-acting protection.

The introduction of the semiconductor devices presented a new situation to the fuse manufacturers in that previously the fuse elements were used to protect metallic conductors, as for example the winding of a motor, and these had similar properties to the metallic elements making it relatively easy to match the characteristics. The matching of the behaviour of metal elements with the semiconductors was clearly more difficult because of their basic differences. Nevertheless, the great progress which has been made was described by Howe and Newbery [36], and the present situation is outlined next.

7.8.1   Protection requirements

Semiconductor devices are of relatively small size for their ratings, as can be judged from Figure 7.23, and their poor overload capacities are illustrated by the example shown in Figure 7.24. The limitation on the periods for which these devices can carry over-currents arises because of their small masses and low thermal capacities, coupled with the high resistivities of the semiconducting materials which leads to internal heating. As an example, a particular stud-type thyristor rated at 70 A mean and 110 A RMS can carry six times the latter value for only 10 ms. In addition, with thyristors there is a limit to the permissible rate of change of current which each device may carry without damage occurring because of localised heating.

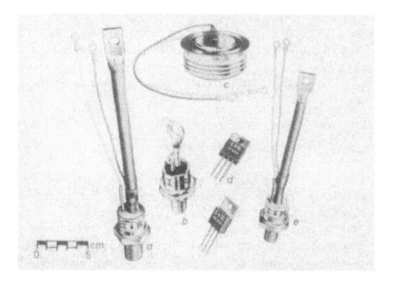

Figure 7.23   Range of thyristors
Where
a   170 A mean
b   25 A mean
c   550 A mean
d   5 A mean
e   80 A mean

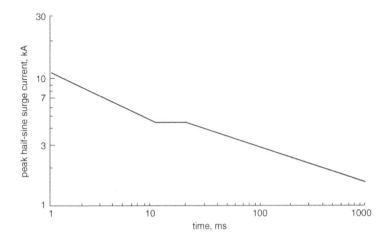

Figure 7.24   Non-repetitive half-cycle fault-current rating of a typical 170 A mean, 1200 V thyristor

Semiconductor devices are also susceptible to damage if over-voltages are applied in the reverse direction, and in some cases transient peak voltages of only twice the peak value of the rated steady-state voltage must not be exceeded.

It is usual for several semiconductor devices to be present in one piece of equipment, such as an invertor or rectifier. The protection of such equipments is examined in some detail later in this chapter, but in all cases the protective equipment should ideally ensure that the following conditions are met:

(a)   In the event of a semiconductor device becoming defective, interruption should be effected quickly enough to prevent damage to other devices. In this connection, experience has shown that semiconductors usually become effective short circuits when they fail and as a result large current flows can result.

(b)   For other faults in the equipment, interruption should take place before there is consequential damage to the semiconductor devices.

(c)   Potentially damaging over-currents should be cleared before devices are damaged.

(d)   Operation of the protective equipment should not cause unacceptably high over-voltages to be impressed on any of the semiconductor devices.

7.8.2   Basic protective arrangements

Cartridge fuses, because of their current-limiting property which enables them to restrict the let-through I^2t values as well as restricting the peak levels of current during short circuits, are basically suitable for protecting semiconductors. Conventional designs are not suitable, however, and special fuselinks, now generally known as semiconductor fuselinks, have been developed. They restrict the let-through I^2t values to even lower levels than do normal fuselinks, and their arc voltages are also lower.

These fuselinks are now so well established that international and national specifications, *IEC 60269-4* (*BS88-4*) have been issued to cover their overall performance
requirements.

In high-power applications, where the cost of the individual semiconductor
devices is considerable, the fuselinks are used in conjunction with other over-current-
protective devices, the former providing protection against only the very high fault
currents. In such situations the fuse cost might be about 10 per cent of that of the
device it protects. The protection for the smaller over-currents may, for example,
be provided by AC and/or DC circuit breakers or for thyristors by suppressing their
gate pulses if the currents or the temperatures of the heat sinks reach critical levels.
A typical protection scheme consists of an AC circuit breaker in the main supply
circuit and fuses in series with the individual semiconductor devices. The circuit
breaker protects against the low-current faults which may be allowed to persist for
relatively long periods (greater than 1 s) and the fuses protect against high-current
faults which must be cleared quickly. This is shown graphically in Figure 7.25 and it
will be realised that the fuselinks do not need to be able to interrupt currents down to
their minimum-fusing level.

Fuselinks for the protection of semiconductor devices in accordance with
IEC 60269-4 have traditionally been 'partial range' or 'back-up' fuses. They are given
the designation or utilisation category of 'aR'. These have a standardised 'minimum
breaking current' where the fuse elements melt in a time of at least 30 s.

As protection schemes and practices have developed, there is a growing need for
fuselinks with 'full range' breaking capacity. An example being to place fuselinks at
the head of the circuit from the supply rather than in the converter cubicle. In this

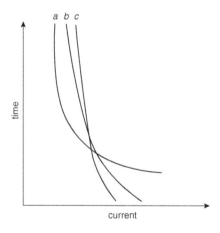

Figure 7.25   *Fuselink providing short-circuit protection*
         Where
         a   circuit breaker characteristic
         b   diode or thyristor characteristic
         c   fuselink characteristic

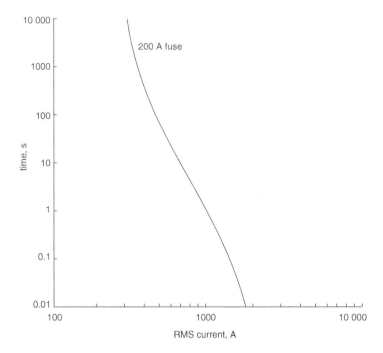

Figure 7.26   Time/current characteristic of 200 A semiconductor fuselink

case the fuselink needs to give protection to the associated cable, in addition to the power semiconductors in the converter equipment. In 2002, two additional full range breaking capacity classifications were introduced into *IEC 60269-4* namely 'gR' and 'gS'. 'gR' are optimised to low I^2t and 'gS' optimised to low power dissipation, for example giving compatibility with standardised fuse bases and fuse combination units.

It can be seen from Figure 7.25 that both the circuit breaker and fuse have inverse time/current characteristics and each has a minimum current level below which it will not operate. A typical fuselink characteristic is shown more clearly in Figure 7.26. In lower-power applications, where the rated currents are less than 10 A, the protection is usually provided wholly by fuses which ensure that fault currents are interrupted before damage can be caused to wiring or other circuit components and physical destruction of the device is also prevented. The fuse may not, however, prevent damage to the semiconductor junction.

To enable the appropriate fuselinks to be chosen, the manufacturers of semiconductor devices have determined and published very complete data in recent years. Characteristics which show the permissible times for which low over-currents may flow in components are available, together with the I^2t values which can be withstood at high current levels and the permitted maximum rates of change of current (di/dt). In addition, although it is strictly excessive energy input which causes damage, manufacturers quote peak instantaneous currents which must not be exceeded and the peak reverse voltages which may be permitted across non-conducting thyristors and diodes are also quoted.

7.8.3   Co-ordination of fuselinks and semiconductor devices

As stated earlier, the I^2t let-through by a fuselink and the voltage of the system in which it is used are interdependent. This is because the effect of a given burnback and arc voltage is reduced when the system supply voltage is increased and therefore the time needed for the current to reach zero is also increased. To enable this factor to be taken into account, some fuse manufacturers provide characteristics, of the form shown in Figure 7.27, relating the I^2t to system voltage.

The I^2t withstand levels of semiconductor thyristors and diodes are established from standardised surge withstand tests of half-cycle duration, 10 ms at 50 Hz or 8·3 ms at 60 Hz. These tests are undertaken with the semiconductor device initially at maximum junction temperature, typically 125°C, and after the pulse test it must withstand full reverse voltage. Since there is a statistical spread in the characteristics, semiconductor manufacturers will quote an I^2t withstand value which will cover all their acceptable products, i.e. it will be the statistical minimum withstand value.

Fuselinks are tested in AC circuits with the most adverse circuit conditions that are likely to apply in service: low power factor, controlled point on voltage wave for the

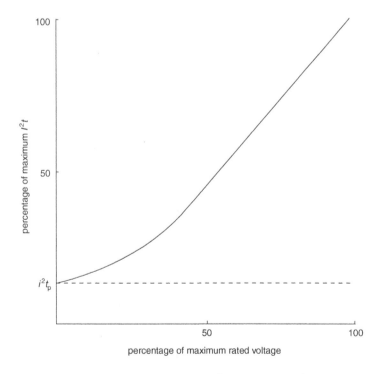

Figure 7.27   Effect of system voltage on total I^2t let-through; $i^2t_p = pre$-arcing I^2t

initiation of the short circuit and single-phase circuits with full line-to-line voltage. The fuselinks are tested with no preloading, i.e. at ambient temperature, typically 25°C. The I^2t let-through of the fuselink is thus the maximum statistical value.

Modern semiconductor fuselinks have I^2t let-through values which do not significantly increase with prospective current at rated voltage. At lower voltages there is an appreciable reduction in I^2t let-through following the reduction in pre-arcing I^2t at shorter times.

To simplify the presentation of information on I^2t let-through, a single maximum value is given by manufacturers at rated voltage with additional values at selected popular voltages. In addition, values at intermediate voltages can be obtained from graphical extrapolation as shown in Figure 7.27.

A simple and effective means of co-ordinating the short-circuit protection of power-semiconductor thyristors and diodes with associated fuselinks can be achieved by comparing the half-cycle withstand with the I^2t let-through by the fuselink. From the foregoing considerations this does give a number of safety factors and in some cases semiconductor manufacturers quote higher I^2t-withstand values if the device does not have to withstand full reverse voltage and the junction is at ambient temperature. The former aspect covers the situation of a fuselink in series with a semiconductor device, and when the fuse operates it becomes open circuit and the device does not have to withstand the voltage. In addition to the above, in many circuits two fuselinks in series clear the fault current, thus assisting arc interruption and reducing the let-through I^2t.

The above simple and effective means of co-ordination covers the majority of applications. However, in critical applications some account may need to be taken of the reduction in I^2t of the semiconductor device with pulse duration. Two graphical methods of co-ordination are shown in Figures 7.28 and 7.29.

7.8.3.1 Effects of the inductance to resistance ratio of a circuit during fault conditions

The I^2t let-through by a fuselink in AC applications is affected not only by the prospective current, but also by the inductance to resistance ratio of the protected circuit during fault conditions. The higher the L/R ratio, the lower the stored energy in the inductance which will tend to reduce the I^2t let-through. However, with a higher value of resistance, it is possible, if the point on voltage wave is so selected, that a greater rate of rise of fault current can be obtained at a lower L/R ratio. This will produce a higher cut-off current which will tend to increase the I^2t let-through. The net effect is that there may not be a significant reduction in I^2t at lower L/R ratios.

7.8.3.2 Effect of ambient temperature

As stated earlier, fuselink characteristics, unless otherwise stated, are obtained at an ambient temperature of 25°C. Increasing the ambient temperature lowers the minimum fusing current because of the reduced heat dissipation and therefore to maintain

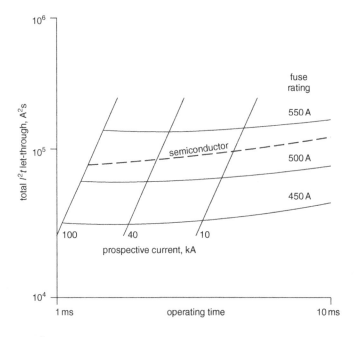

Figure 7.28 I^2t co-ordination

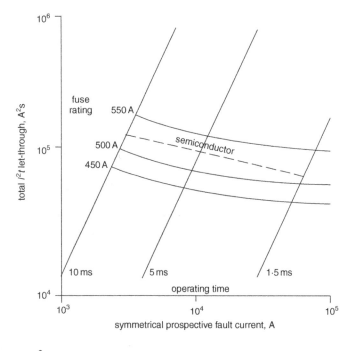

Figure 7.29 I^2t co-ordination

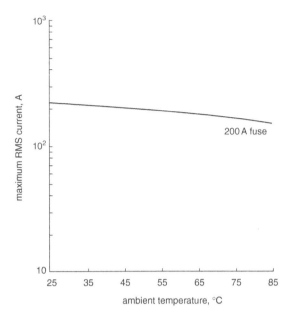

Figure 7.30   Changes in fuselink current rating with ambient temperature

an adequate fusing factor, the rated current must be correspondingly reduced. Relationships like that shown in Figure 7.30 are provided by manufacturers to enable the correct derating to be determined.

7.8.3.3   Effects of forced cooling

To achieve the maximum ratings of diodes and thyristors, it is often necessary to force-air-cool them. If the protecting fuselinks are also cooled by the air stream, then fuselinks with slightly lower ratings than would normally have been chosen may be used, as indicated in Figure 7.31. For example, if calculations indicate that a 200 A fuselink should be used in a circuit operating without air cooling in an ambient temperature of 20°C, then a 175 A fuselink could be used if an airflow velocity of 2 m/s or more is provided.

Figure 7.31 may be used in conjunction with Figure 7.30 if the still-air temperature is greater than 25°C.

When forced-air-cooling is used, the fuselink body and end caps are kept quite cool, but the temperature of the element, at the uprated current, is similar to its value at the rated current without cooling. Thus, if fuselinks are rerated to levels higher than those shown in Figure 7.31, there is a risk that they will maloperate.

7.8.3.4   Connections

Semiconductor fuselinks are usually connected to busbars and up to 75 per cent of the heat generated in a fuselink is dissipated to them through the end connections.

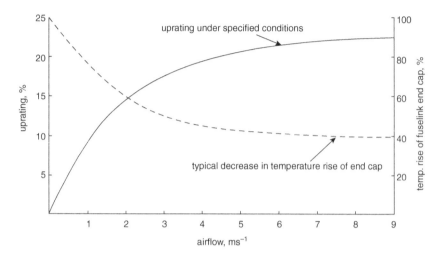

Figure 7.31   *Uprating of semiconductor fuselink with forced-air-cooling and decrease in the temperature of end cap*

As fuselink characteristics are determined using a standard test rig which, for example, has busbars of a 25·4 mm × 6·35 mm cross-section for a 200 A fuselink, the fuselink's ratings must be changed if the cross-sectional and surface areas of the busbars are different from those used by the manufacturers for their tests. The characteristics for assigning the appropriate ratings are shown in Figure 7.32. As an example, if a 200 A fuselink is connected to busbars 12·7 mm × 9·5 mm, the cross-sectional area is only 75 per cent of that used by the manufacturers and the surface area is only 70 per cent. The fuselink must therefore be derated to 96 per cent of 200 A, i.e. 192 A.

7.8.4   Cyclic loading of semiconductor fuselinks

Equipment which incorporates semiconductor devices and, thus, semiconductor fuselinks is often subjected to repetitive overloads. Under these conditions, the temperatures of the fuselink elements rise and, with large overloads, the element temperatures may approach the melting point of silver and fatigue may result and lead to incorrect fuselink operation. Small repetitive over-currents cause the mechanical stresses in an element to change continually and this also fatigues the element and so may cause premature operation of the fuselink. Experience has shown that incorrect operation can be avoided if fuselinks used in circuits subjected to short repetitive overloads have a nominal melting RMS current, at a time equal to the overload duration, of at least twice the overload current. As an example, if an overload of 450 A and 5 s duration continually re-occurs, then the fuselink should have a nominal melting current of at least 900 A at 5 s.

When overloads of longer than a few minutes re-occur regularly then the rated current of the fuselink should be at least equal to the overload current.

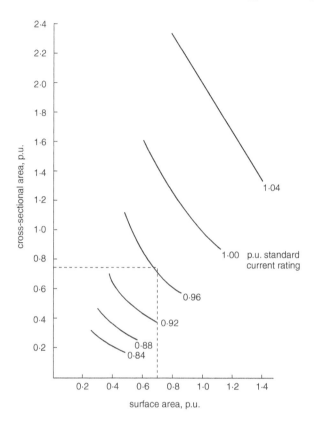

Figure 7.32   Variation   of   current   rating   with   conductor   cross-section   and surface area

7.8.5   Application of fuselinks to equipment incorporating semiconductor devices

The fuselinks have two main functions:

(*a*)   To so limit the magnitudes and directions of the currents which can flow through healthy devices that damage does not occur.

(*b*)   To protect the remainder of the circuit when a semiconductor device becomes faulty.

As stated earlier, an operating fuselink must:

(*i*)   limit the peak current and the I^2t let-through to devices conducting at the time of the fault

(*ii*)   restrict the voltage impressed across devices which are blocking when the fault occurs.

To show how these objectives are achieved in practice, a number of circuits are considered below. Rectifier equipments are examined first, particular attention being paid to the single-phase, full-wave bridge rectifier and large multiphase rectifiers with several parallel paths.

7.8.5.1 Protection of rectifiers

One possible fault which can occur is a short circuit in the load supplied by a bridge rectifier. In the event of such a fault a very large current will flow through two of the semiconductors (say D_1 and D_3) and the associated fuselinks (F_1 and F_3) as shown in Figure 7.33. These fuselinks will start to operate and limit the current, as shown in Figure 7.34, to thus meet the requirement that the fuselink limits the peak current and I^2t let-through. As the fuselinks operate, the voltages across the blocking diodes (D_2 and D_4) drop initially and then rise to amplitudes which may be greater than the supply voltage. These voltages, which are approximately equal to the fuse arc voltages, can destroy the diodes if the voltage amplitude exceeds the maximum reverse potential for the rectifiers.

   If the load fault is solid, then fuses F_2 and F_4 will operate during the next voltage half-cycle as illustrated by the waveforms.

   To achieve proper co-ordination and prevent healthy devices from being destroyed, the fuselinks selected must ensure that the peak current and the I^2t withstand of the devices are greater than the maximum let-through by the fuselinks and also that the peak reverse voltage of the rectifiers is greater than the maximum fuse arc voltage.

   Another type of fault that can occur in a full-wave bridge rectifier is for one of the semiconducting devices to short-circuit while it is reverse biased. As an example,

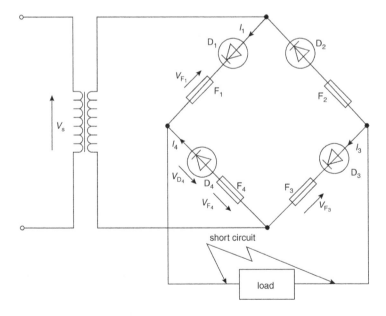

Figure 7.33   Single-phase full-wave bridge rectifier with short-circuited load

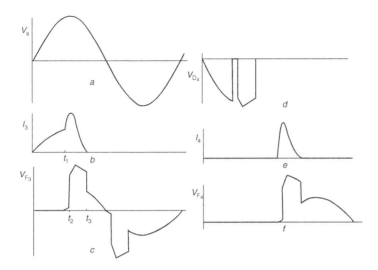

Figure 7.34   Oscillograms recorded during the clearance of the fault

Where

a   supply voltage

b   current through diode D$_3$ (t_1 is time of fault initiation)

c   voltage across fuse F$_3$ (t_2 is start of arcing, t_3 fuse clears fault)

d   voltage across diode D$_4$

e   current through diode D$_4$

f   voltage across fuse F$_4$

the behaviour when the diode D$_2$ in the rectifier shown in Figure 7.35 short-circuits during one of its non-conducting periods is considered. Immediately, the currents I_2 and I_3 will increase rapidly, as shown in Figure 7.36, and fuses F$_2$ and F$_3$ will start to operate. As diode D$_1$ will still be conducting the majority of the fuse voltage V_{F_2} will be impressed across the diode D$_4$, which is blocking.

As with a load fault, the fuselink must limit the peak current, I^2t let-through and arc voltage to values which will avoid destruction of the three healthy devices. The method of co-ordination is therefore identical to that described above.

Multiphase high-current rectifiers are used extensively in industries such as refining, electroplating, chemical plants and steel manufacture. Because of the very high current ratings of the equipment, it is necessary to operate semiconductor devices in parallel. An additional requirement is that this expensive high-power plant is not taken out of service, other than for routine maintenance, and it is therefore necessary to use more than the minimum number of devices in parallel (i.e. there is a built-in redundancy) to ensure that production is maintained. A typical rectifying plant is illustrated in Figure 7.37. This type of circuit is normally protected by fuselinks in series with the individual devices and circuit breakers in the AC lines but fuselinks are occasionally used throughout. Only the first case will be considered here. It is clear from the circuit diagram that all fault currents will be fed through the circuit breaker

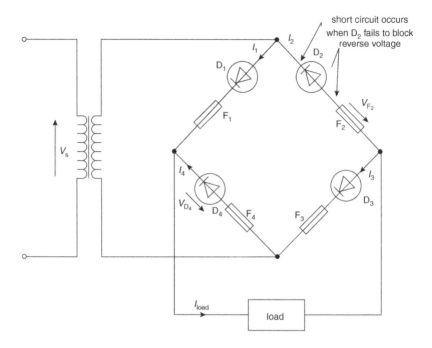

Figure 7.35   Single-phase   full-wave   bridge   rectifier   with   device   which   has an internal short circuit

but, as stated earlier, it only offers protection against the relatively small-magnitude overloads of long duration whereas the fuselinks respond to the short-period high over-currents. As the currents associated with load faults tend to be of small magnitude, they are cleared by the circuit breaker. The fuselinks therefore protect against internal faults which cause large currents to flow.

When a device short-circuits internally, the fuselink connected in series with it must clear before the other semiconductors are damaged and the faulty device explodes. To do this, the following conditions must be satisfied.

(*a*)   The total I^2t let-through of the fuselink in series with the faulty device must be less than the explosion rating of the device.

(*b*)   The total I^2t let-through by the fuselinks in series with the fault, but protecting healthy devices, must be smaller than the value quoted by the manufacturer for the pre-arcing I^2t of these fuselinks. This ensures that, after the fault has been cleared, the fuselinks protecting healthy devices are not fatigued and will not operate unnecessarily at some later time.

(*c*)   The peak arc voltage of the operating fuselink must be less than the maximum reverse voltage for the devices.

To examine how these requirements may be met, the situation is considered which arises when a diode fails as it is blocking half the reverse peak voltage. Under this condition the fault current feeds through only one arm of the bridge, as illustrated in

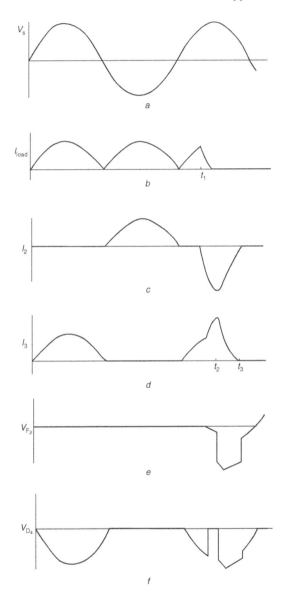

Figure 7.36   *Oscillograms recorded during clearance of fault*

Where

a   supply voltage

b   load current (t_1 is time of fault initiation)

c   current through diode D_3

d   current through diode D_3 (t_2 start of arcing, t_3 fuse clears fault)

e   voltage across fuse F_2

f   voltage across diode D_4

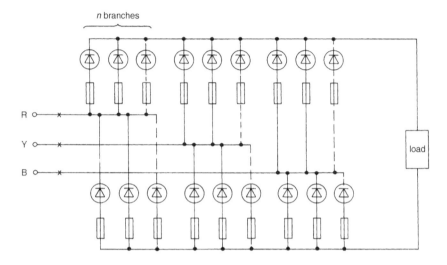

Figure 7.37   Three-phase bridge rectifier with parallel paths

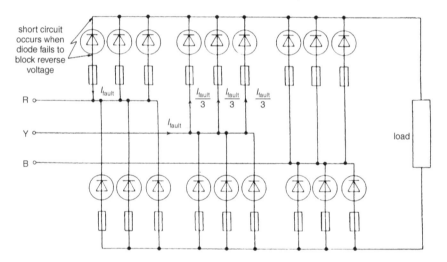

Figure 7.38   Three-phase, parallel-path, bridge rectifier with a device which has an internal short circuit. For clarity only fault currents are shown

Figure 7.38. The I^2t let-through by each of the fuselinks in the healthy conducting branch is

$$\frac{I_F^2 t_c}{n^2}$$

where I_F is the fault current, n is the number of diodes in parallel and t_c is the clearance time of fuselink in series with the faulty device.

This quantity must be considerably less than the pre-arcing I^2t of these fuselinks and the I^2t withstand of the diodes if none of these components are to be damaged while the fuselink in series with the faulty device is operating, i.e.

$$\frac{I_F^2 t_c}{n^2} < I^2t \quad \text{pre-arc} \; < I^2t \; \text{diode withstand}$$

Four or more devices must be used in parallel for this condition to be achieved.

7.8.6 Protection of DC thyristor drives

The DC thyristor drive has traditionally been a popular means of achieving a variable speed motor drive. In most cases, the protection scheme includes fast-acting semiconductor fuselinks for the protection and isolation of the power thyristors in the very short time protection region associated with high fault currents. Information on fusing DC drives is given in a technical report *IEC 60146-6* Application Guide for the Protection of Semiconductor Converters against Over-current by Fuses. It covers line commutated converters in single-way and double-way connections and includes non-regenerative and regenerative drives. An example will be taken of a typical DC variable speed drive using a fully controlled three-phase bridge, as illustrated in Figures 7.39 and 7.40.

It is usual practice to protect the thyristors against high short-circuit currents caused by internal or external faults. In a typical non-regenerative drive, the fuselinks are usually connected in series with the thyristors to give the best possible protection.

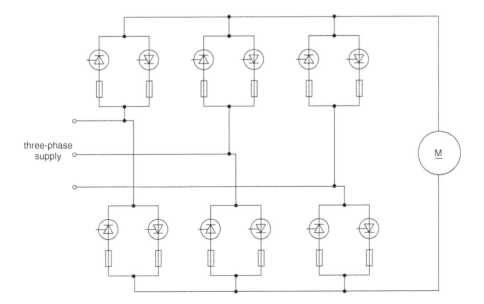

Figure 7.39   Basic thyristor-convertor drive circuit

Figure 7.40   A 1.1 MW thyristor-convertor drive

For drives which are subjected to repetitive load-duty cycles, account must be taken of the pulsed-type loading, as explained in Section 7.8.5.

If the drive is regenerative, many types of fault may occur which make fuse protection difficult. In some cases, the fuselinks have to clear DC faults. As outlined earlier, the absence of voltage zeros and possible long-time constants associated with DC faults make fuselink clearance much more difficult than it is with alternating currents, and an adequate DC fuse-voltage rating must be chosen. To illustrate the situation clearly, two common fault conditions are considered below.

It is possible during regeneration, when thyristor ThB2 in Figure 7.41 is conducting normally and ThB1 is blocking, that the latter may receive a spurious gate pulse and be accidentally triggered. The terminals of the DC machine will immediately be short-circuited, and the two fuselinks in series with the thyristors will share the duty of clearing the fault. It is unlikely that the two fuse arc voltages will be equal, as differences, as large as 4 : 1 ratio, have been recorded. Because of this, the direct voltage of fuses should be greater than the maximum DC regenerative voltage.

A second fault condition can occur when there is a low-impedance path through the AC supply and DC machine. This would occur if thyristor ThB2 were conducting normally and ThR1 accidentally switched on, as shown in Figure 7.42. In this case, the fuselinks would have to clear the sum of the alternating and direct voltages. Fortunately, the peak alternating voltage is always greater than the direct voltage and natural voltage zeros arise. The RMS alternating-voltage rating of the fuses should then be at least

$$V_{\text{AC(RMS)}} + \frac{V_{\text{DC}}}{\sqrt{2}}$$

Occasionally, a high-speed DC circuit breaker is connected between the thyristor convertor and the DC machine but, unfortunately, it is difficult to co-ordinate the

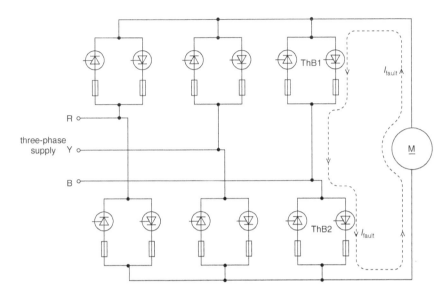

Figure 7.41   Thyristor-convertor drive circuit with DC fault

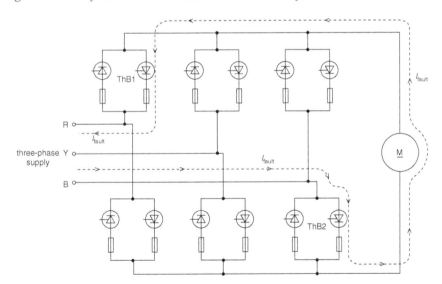

Figure 7.42   Thyristor-convertor drive circuit with fault current flowing through AC supply and DC machine

fuselink and circuit-breaker protection in this configuration. For example, if the AC supply is lost, it is necessary for the DC breaker to operate before the fuselink elements start to melt. Normally, circuit designers would choose fuselinks which had a small I^2t let-through, in order that protection against the high over-currents associated with the internal short-circuit of semiconductor devices could be provided. But, unfortunately, in this situation this cannot be done as it is necessary to maintain discrimination between the fuselink and circuit-breaker protection, and so fuselinks with a high pre-arcing I^2t let-through must be used.

7.8.7 Protection of inverters

Inverters are another group of circuits which require fuse protection.

In inverters, one example of which is shown in Figure 7.43, the power flow is from the DC supply to the AC output. Commutation equipment is provided (indicated by thin lines on the diagram) to force the thyristors to switch off at the correct times and produce an alternating voltage at the output. It is failure of this commutation plant which can lead to two thyristors in the same phase conducting simultaneously and short-circuiting the DC supply.

A fault can also occur when a thyristor, which should block current in the forward direction, is accidentally switched on by a spurious gate pulse. Although the causes of these two types of fault are different, the effects can be similar, so these faults are often referred to jointly as 'DC shoot-throughs'.

To analyse faults associated with inverter circuits, the situation when the DC supply is short-circuited by one of the fault conditions described above is now examined. The smoothing capacitor begins to discharge and a very large-amplitude

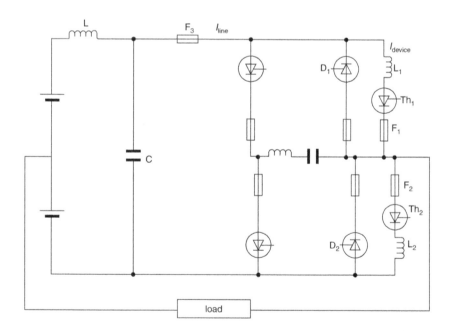

Figure 7.43   *Auxiliary impulse-commutated inverter, commonly called McMurray inverter. For clarity only one phase is shown*

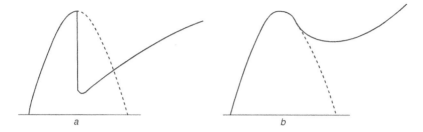

Figure 7.44   Fuse-current waveforms during inverter fault clearance
       a   line fuse
       b   device fuse

oscillatory current starts to flow through the thyristors Th_1 and Th_2 and the small lead inductances L_1 and L_2. The current rises to a maximum fault level in between 1 and 3 ms. Immediately the capacitor voltage reaches zero and the current attains its maximum value, the current is diverted through the reverse diodes D_1 and D_2 producing a rapid drop in capacitor current and relatively quick decay of thyristor current. Current will then start to flow from the supply producing a second, but relatively slow, increase in fault current, as illustrated in Figure 7.44.

To protect the inverter, the fuselinks connected in series with the thyristors are usually chosen so that they will melt before the capacitor discharges. The fuselinks then have to clear a relatively low voltage in a circuit of low inductance. This is a relatively light duty and there is little arcing. The thyristor current is rapidly reduced to a low value but, owing to the increase in fuselink resistance, the time constant of the circuit increases and the thyristor current tends to persist for several milliseconds. During this time, the capacitor begins to recharge and the current starts to increase again. To ensure that this current does not persist for too long, care is required in the selection of the fuselink.

The alternative approach is to design a circuit which can cope with the initial high-current surges and choose a fuselink which will operate after the current is diverted through the diode. The circuit magnitude and time constant can be calculated, and then the most appropriate fuselink may be determined. When making this calculation, the effects of capacitance and circuit inductance should not be ignored, as they affect fuselink clearance.

The inverter circuit described was relatively simple. The protection of more complex networks requires extreme care.

7.8.8   Protection of power transistors

A factor which must be taken into account when using thyristors is that they do not switch off automatically. For example, in inverter circuits, expensive commutation equipment must be provided to ensure that conduction stops at the correct moment. Transistors, however, only conduct when a base current flows. As a result, high-power transistors have been developed, which may supersede thyristors. Power transistors which can operate at voltages up to 800 V and collector currents of 250 A are now available. However, the voltage rating is reduced as the collector-current rating is increased, so that, at collector-current ratings of 400 A, the voltage rating is typically 400 V.

In high-power applications, transistors are operated in the saturated condition (i.e. the collector–emitter voltage is low, and the collector and base currents are high). With this arrangement, the internal power loss is kept to a minimum. The disadvantage of operating in the saturated region is that, immediately a fault occurs which reduces the load impedance, the collector current and collector–emitter voltage rise simultaneously. The voltage appears across the transistor, since the transistor will support that voltage in the active state, and the current rises to a value dependent upon the available base drive and gain of the transistor. The simultaneous increase in collector current and collector–emitter voltage causes an increase in the internal heat dissipation in the transistor. Theoretically, to prevent transistors being operated outside their saturation regions, transistor circuits could be designed so that, under fault conditions, base currents rise to maintain the saturated operation sufficiently long (up to 10 ms) for semiconductor fuselinks to operate. Unfortunately, this is not a practical proposition because many higher-power transistors have current gains of only two or three at high collector currents so that increasing the base currents during fault conditions would, in itself, destroy the devices. The more usual approach is to detect over-currents rapidly and remove base drive to effect turnoff. Alternatively, circuits have been designed with automatic current limiting, current feedback, over-voltage 'crowbars', foldback output characteristics and temperature-sensitive trips. In all these cases, the power supply to the system is protected by fuses, but the individual transistors are left unprotected.

The fuse chosen still needs to be a fast-acting type to protect the power transistors from case rupture during the passage of high currents which could in turn cause consequential damage to the convertor.

The increasing use of AC drives, utilising the robust squirrel cage rotor, over the DC drive, with its associated expensive commutator, has been brought about by the development of the power transistor. AC drive inverters have changed considerably since around the mid-1990s with thyristor technology being replaced with insulated gate bipolar transistor (IGBT) technology.

This has greatly helped improve the efficiency and handling of non-linear loads and provides extremely low voltage distortion. With the older thyristor technology, the voltage waveform peaks were frequently flattened. This does not occur with IGBTs. Very high switching speeds are also possible, as is reduced on-forward drop and increased short-circuit withstand capability. All this means that AC motor drives using IGBTs provide a much better and more tightly controlled performance. IGBTs have also been applied with equal success to other inverter applications such as for static uninterruptible power supplies (UPS).

A further advantage is that use of IGBTs reduces inverter size and allows repairs to be made more easily. Moreover, digitally controlled inverter design and the growth in remote monitoring now allow control and monitoring of drives remotely. For the above reasons, the use of IGBTs in AC in motor drives, UPS and other systems is now very common.

Nowadays, IGBTs are readily available with ratings of 3·3 kV and 1200 A RMS and can thus cover the bulk of the AC drives market.

IGBTs are typically operated at frequencies around 10 kHz which puts some additional requirements on circuit and component design and selection, including fuses. As described earlier in this section, electronic protection against overloads and short circuits is normally provided, but back-up fuse protection is still needed to ensure safety in the event of failures of these systems or the device itself. A weak link in IGBTs is the internal emitter wire connections and the explosion rating of the IGBT is often governed by the I^2t to melt these connecting wires. This can be calculated from the cross-section, A in mm^2, of the wire using the following 'adiabatic' formulae:

$$I^2t = 100\,000A^2 \qquad \text{for copper}$$

$$I^2t = 80\,000A^2 \qquad \text{for silver}$$

As described in Section 7.8.7 the fault current waveform is associated with the discharge of the capacitor and the fuse needs to melt before the first peak of the fault current. If this is done, then the interruption of the current will be rapid against the decreasing voltage from the discharge capacitor. The arcing I^2t will therefore be a similar value to the pre-arcing I^2t. Reference to I^2t let-through and voltage rating of the fuse related to the DC link voltage should be established, taking into account the C, R and L of the fault circuit.

In IGBT inverters, an additional de-rating factor may be necessary due to skin and proximity effects – see Section 2.4. Busbars are also subject to additional heating and high frequency and if they are not sized properly, this may also require a further reduction in the continuous current rating, due to the heating effects at the ends of the fuse.

Since IGBT circuits are switched at high frequencies, the switching di/dt is high and the circuit inductance must therefore be kept low to minimise transient overvoltages. The additional inductance covered by the insertion of a fuse is primarily due to change in the circuit loop. The fuse internal self-inductance can be ignored.

Special geometry designs are available for IGBT protection as shown in Figure 7.45. These are often a flat-pack design with a single element. However, the fuse is only part of the loop and it is necessary to consider the geometry of the fuse and circuit together.

7.8.9 Situations where there are high surge currents of short duration

The protection of circuits in which high surge currents flow is particularly difficult. A common example is resistance welding. This is a case of cyclic loading where, under normal operating conditions, the load impedance is very low and it is therefore difficult for fuselinks to discriminate between short-circuit faults and normal operating conditions. Thus, in a few cases, it is only possible for the fuselinks to prevent damage to the wiring and other circuit components and to avoid the physical destruction of the devices, but not to protect the semiconductor junctions.

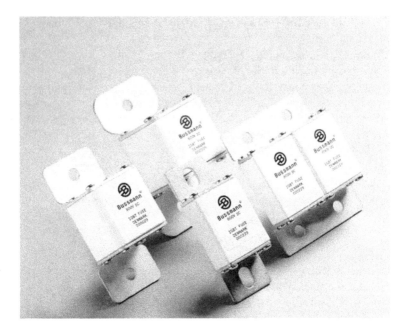

Figure 7.45   'Flat' fuselinks for IGBT protection

There is also a serious difficulty in the protection of triac- or thyristor-controlled tungsten–halogen lamps. When the lamps are switched on, large surge currents flow unless the dimmer circuits have facilities automatically to control the position of the gate pulses with respect to the voltage wave during the first few cycles, to provide what is commonly called a 'soft start'. Where this facility is not provided, it is necessary to use a semiconductor and fuselink which have continuous ratings well above the normal lamp current to cope with the inrush current. Otherwise, there is a risk of the fuse deteriorating with each starting surge and eventually failing as explained earlier.

Sometimes the extra cost of using a semiconductor and fuselink well above the normal lamp current is not warranted. In this situation, no semiconductor protection is provided. High-breaking-capacity industrial fuselinks are fitted, to prevent the semiconductor from exploding and the wiring from being destroyed when the lamp becomes short-circuited. Industrial fuselinks are more robust than semiconductor fuselinks and are more suitable for withstanding current surges.

7.8.10   Special applications

There are many situations where semiconductor fuselinks are required but the standard design is unsuitable. For example, in brushless alternators, the field winding rectifiers and fuselinks are mounted on the rotor as shown in Figure 7.46. These revolve at speeds up to 3000 rev/min and the fuselinks are therefore subjected to both vibration and large forces and so require careful mechanical design.

Figure 7.46   Fuselinks mounted on brushless alternator

In many situations, the complete protection of semiconductor devices is difficult or is not economically justified, for example, in circuits with either low-power thyristors and diodes or power transistors. But, as protection against physical destruction must always be provided, a demand for fuselinks will remain, although there may be a transfer from the semiconductor types to the more robust industrial designs.

At present there are difficulties in co-ordinating high-power water-cooled devices with fuselinks, which means that sometimes the semiconductors have to be operated below normal rating. To alleviate this, fuselinks which have lower I^2t let-through are being developed.

The continual development of inverter circuits is increasing the demand for fuselinks which satisfactorily clear DC faults. Although much progress has been made in this field, improvement in fuselink design is still needed. Over the last decade there has been a continual reduction in the cost of power semiconductors, but it now appears that prices have stabilised in real terms. The role of the fuselink as an economical means of protection of power semiconductor devices is likely to remain in the foreseeable future.

7.9   Protection against electric shock

In low-voltage electrical installations, protection against electric shock is dealt with in *IEC 60364-4-41*. This IEC standard covers protection against direct and indirect contact.

Direct contact is when contact (by persons or livestock) is made directly with a live part and which is likely to cause current to flow through a body to the injury, perhaps fatal, of that person or animal.

Indirect contact is when contact is made with an exposed conductive part which is not live under normal conditions, but which may become live under earth fault conditions. This hazard arises when the protective measures against direct contact have ceased to be effective where, for example, there is failure of insulation.

7.9.1   Protection against direct contact

IEC 60364-4-41 includes the following basic protection measures:

- Protection by insulation of live parts.
- Protection by a barrier or an enclosure.
- Protection by obstacles.
- Protection by placing out of reach.

Examples of the protection by a barrier or enclosure for fuses and fused equipment are illustrated in Chapter 4. It is a requirement that live parts are to be inside enclosures or behind barriers, which provide a degree of protection to at least IP2X or IPXXB, (*IEC 60529*) preventing ingress of 12 mm diameter objects such as fingers. In addition, the top surface of an item of equipment which is also readily accessible is required to have an IP4X or IPXXD degree of protection – preventing ingress of objects of 1 mm diameter such as small wires, tools, insects and general debris.

7.9.2   Protection against indirect contact

IEC 60364-4-41 includes the following basic protection measures:

- Protection by earthed equipotential bonding and automatic disconnection of supply.
- Protection by class II equipment (additional insulation but no means of protective earthing provided) or equivalent insulation.
- Protection by non-conducting location.
- Protection by earth free local equipotential bonding.
- Protection by electrical separation.

Protection against indirect contact can therefore be provided by earthed equipotential bonding and the automatic disconnection of supply. Fuselinks can provide the automatic disconnection of supply and is based on fuselink operation, which disconnects the supply circuit. In the event of a fault between a live part and an exposed conductor in the circuit or equipment a prospective touch voltage must not persist long enough to cause physiological effects to a person. A disconnecting time of 5 s is allowed under certain circumstances and in some cases shorter disconnecting times are required, for example, 0·4 s.

To determine the condition for disconnecting the supply, the required time for disconnecting shall be considered, according to the type of system earthing and the environment specified in *IEC 60364-4-413*.

There are three basic low-voltage distribution systems, TN, TT and IT, depending upon the relationship of the source and of exposed conductive parts of the installation, to earth. These three types of system are identified as follows:

- TN system, a system having one or more parts of the source of energy directly earthed, the exposed conductive parts being connected to that part by protective conductors.
- TT system, a system having one part of the source of energy directly earthed, the exposed conductive parts of the installation being connected to earth electrodes electrically independent of the earth electrodes of the source.
- IT system, a system having no direct connection between the live parts and earth, the exposed conductive parts of the electrical installation being earthed.

IEC 60364-4-413 specifies the requirements for the automatic disconnection of supply for these systems. An example of the requirements for a distribution circuit in a TN system is as follows.

A disconnection time of 5 s is permitted for a distribution circuit. The maximum impedance of the fault loop, Z_s, is calculated from the following formulae for a given fuselink:

$$Z_s \leq \frac{U_0}{I_a}$$

where U_0 is the nominal voltage to earth and I_a is the current determined from the time current characteristic of the fuselink which causes the operation of the fuselink in 5 s.

For a distribution circuit in a TN system, *BS7671* the British version of *IEC 60364*, gives values of earth loop impedance, Z_s derived from the standardised time–current characteristic zones for gG fuselinks, thus giving generalised values for Z_s.

7.10 Arc flash

At the time of this publication there is a great deal of activity in the USA on the subject of the hazard of arc flash when a fault occurs in electrical equipment.

The installation codes in North America require that equipment be installed in accordance with the way it was listed or certified. The installation codes, however, do not provide guidance for maintenance workers and the product standards do not require testing for arcing faults that might occur when the equipment door is open and a maintenance worker accidentally creates an arcing short circuit. As a result, numerous workers are injured and killed each year while working on energised electrical equipment. An ad hoc working group was formed to help address this situation. The intent of the group was to raise the awareness of electrical workers to the dangers associated with electrical arcs and hopefully reduce the incidents of worker injuries and deaths. The results of the work are given in an IEEE paper [37].

If an arcing fault occurs, then tremendous energy is released in a fraction of a second as illustrated in Figure 7.47.

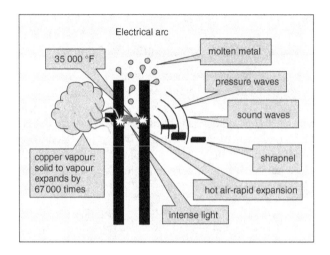

Figure 7.47   Electric arc model

The 2002 edition of *The National Electric Code, NEC*, requires equipment on which work may be required when energised to be labelled, warning of arc flash hazard.

The 2003 edition of *NFPA 70E* (*National Fire Protection Association*) requires a flash hazard analysis to be undertaken, including the determination of the type of protective personal equipment, PPE, needed.

In 1982, a paper by Lee [38] developed a formula for the distance required for various degrees of burns, as related to the available MVA and time of exposure. These formulae were developed for arcs in open air:

$$D_c = (2 \cdot 65 \, \text{MV} \, A_{bf} t)^{1/2}$$

$$D_c = (53 \, \text{MVA} \, t)^{1/2}$$

$$D_f = (1 \cdot 96 \, \text{MVA}_{bf} t)^{1/2}$$

$$D_f = (39 \, \text{MVA} \, t)^{1/2}$$

where D_c is the distance for a just current burn (feet); D_f is the distance for a just fatal burn (feet); MV A_{bf} is the bolted fault; MVA is the transformer MVA, $0 \cdot 75$ MVA and over (for smaller ratings, multiply by $1 \cdot 25$) and t is the time of exposure in seconds.

In 2002, the *IEEE Guide 1584: Performing Arc-Flash Hazard Calculations* was published updating the requirements.

The current-limiting fuse in its current-limiting mode minimises the arc flash hazard. A number of papers have been published [39, 40] and videos produced [41] illustrating how current-limiting fuses dramatically reduce hazards due to electrical arc flashes.

Surprisingly formal 'management' of arc flash hazard until 2003 was restricted to the USA. However, at the 2003 ICEFA Conference (International Conference on Electric Fuses and their Applications) three papers were presented with input from Australia, Stokes and Sweeting [42]; Sloot and Rilsma [43] and UK, Wilkins, Allison and Lang [44].

7.11   Power quality

This topical subject was itemised in the benefits of using fuses given in Section 7.1. Power quality essentially relates to voltage quality, limiting over-voltages and the duration of voltage 'outages', 'dips' or 'sags'. The excellent discrimination properties of fuselinks described in Section 7.2 gives a significant advantage in using fuselinks and particularly in their current-limiting mode of operation.

Wilkins and Chenu [45] outlined the requirements for acceptable power quality and identified the Information Technology Industrial Council (ITIC) curve as the basic standard for all types of equipment and power systems, see Figure 7.48. The over-voltages during fuse operation from pre-arcing (melting) to arcing, are below the threshold given in the ITIC curve and are limited in magnitude in the appropriate IEC fuse standards.

For a bolted short circuit, zero voltage occurs in the faulted circuit and the ITIC curve requires that the fault should be cleared in less than 0·02 s. The current-limiting properties of fuses admirably meets this requirement. It should be realised that fuselinks can become current limiting at relatively low fault currents, see Section 7.1.5, for example, a 32 A gG low-voltage fuselink becomes current limiting at a current as low as 400 A and a 100 A fuselink at 1500 A. This minimises the duration of disturbances in the associated network.

Wilkins and Chenu [45] illustrate a practical example in a typical motor control centre. The operation of a fuse on a bolted short circuit on one motor circuit

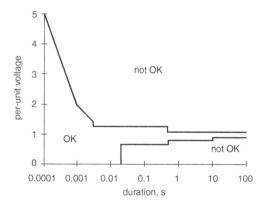

Figure 7.48   The Information Technology Industrial Council curve

gives acceptable low disturbance on other parallel motors with regard to voltage and associated motor performance in the other parallel circuits.

Kojovic *et al.* [46] illustrate examples in medium voltage distribution applications minimising the disturbances in associated power systems when current-limiting fuses are used. Reference 47 shows the advantages of using compact full-range current limiting fuses as a retrofit to expulsion fuses, improving power quality.

Wilkins and Chenu [45] point out that 'in low and medium voltage industrial, commercial and residential networks, current-limiting fuses have been improving power quality for more than 60 years, long before power quality became an important issue'.

Chapter 8

International and national standards

It was mentioned in Chapter 1 that the Electric Lighting Acts, introduced by the British Parliament in the 1880s, gave the UK Board of Trade the responsibility of introducing regulations to secure the safety of the public and ensure a proper and sufficient supply of electrical energy. This led to the production of several documents which specified regulations, and eventually, in 1919, *British Standard 88* was introduced for fuses with rated currents and voltages up to 100 A and 250 V, respectively. Since that time, this standard has been extended and updated and other British Standards have been introduced. Other major countries in which fuses are manufactured have also produced their own national specifications and standards and understandably their requirements have differed from those in the UK. This has been inevitable because of factors such as the different units of length which are used in the various countries, i.e. centimetres and inches, and the different operating voltages which have been adopted.

All national standards are continually being reviewed and revised to meet the changing requirements and also to give wider international acceptance. There is also a desire to have internationally agreed standards which will be used throughout the world and these are being produced by the International Electrotechnical Commission (IEC) and the format and contents of many national standards have been aligned with those of the IEC. Most developing countries, in fact, use IEC standards and thus do not have the complicated infrastructure of national standards.

8.1 Contents of standards

The various standards including those produced by the IEC are not presented in a completely uniform manner but the contents are usually arranged into the following sections:

Scope
Definitions

Standard conditions of operation
Ratings and characteristics
Markings
Standard conditions of construction and test
Type tests
Dimensions
Application guides.

This chapter concentrates mainly on IEC practice and does not attempt to give detailed information which can readily be obtained by referring to the appropriate standards. Each of the sections referred to above is examined and the general aims and items of interest are presented.

8.1.1   Scope

It is clear that any standard must include an initial statement or list of the equipment or devices covered by it. For fuses the requirements vary for the different types which are produced and therefore IEC has produced separate standards for the three categories into which fuses may be divided, namely miniature, low-voltage and high-voltage, these categories being defined as follows.

8.1.1.1   Miniature fuses

These are for the protection of electric appliances, electronic equipment and component parts thereof, normally intended for use indoors.

8.1.1.2   Low-voltage fuses

These are fuses incorporating enclosed fuselinks with rated breaking capacities of not less than 6 kA, intended for protecting power frequency AC circuits of rated voltages not exceeding 1000 V or DC circuits of rated voltages not exceeding 1500 V.

8.1.1.3   High-voltage fuses

These are fuses designed for use outdoors or indoors in AC systems of 50 and 60 Hz and of rated voltages exceeding 1000 V.

8.1.2   Definitions

A 'Glossary of terms' has been included in this Third Edition.

There is an IEC committee, International Electrotechnical Vocabulary (IEV), which is solely concerned with definitions and it is hoped that the terms it agrees will eventually be used everywhere. There are significant variations at present, however, and in particular even the word 'fuse' does not have a single meaning, being used in standards in several countries, including the USA, when referring to the fuselink rather than the complete assembly.

8.1.3 Standard conditions of operation

Because the behaviour of fuses is affected by environmental conditions, it is the practice for standards to state the range of conditions in which fuses will operate satisfactorily.

The following operating conditions are usually included.

8.1.3.1 Ambient temperature

This affects the power which may be dissipated from the surfaces of a fuselink and its fuse holder or mounting for any given element temperature, and thus the operating times at low over-currents are also affected by it. The operating range is not the same for all fuses, the IEC requirements being that low-voltage fuses should be suitable for operation in ambient temperatures between -5 and $40°C$ whilst high-voltage fuses must operate satisfactorily over the wider range of -25 to $40°C$. This is because the latter fuses are often mounted outdoors in exposed positions. Limits are not specified for miniature fuses.

8.1.3.2 Humidity

This condition can affect the insulation levels of fuselinks and their associated parts, and a typical requirement is that satisfactory operation should be obtained in relative humidities up to 50 per cent at $40°C$ and higher levels at lower temperatures.

8.1.3.3 Altitude

This also affects insulation levels and the IEC requirements vary for the different categories of fuses. Low-voltage fuses must be suitable for operation up to 2000 m whilst 1000 m is specified for high-voltage fuses. No value is specified for miniature fuses.

8.1.3.4 Atmosphere

To prevent the possibility of fuses being adversely affected by the surrounding atmosphere, it is usual for standards to contain statements to the effect that the ambient air should not be excessively polluted by dust, smoke, corrosive or flammable gases, vapour or salt.

8.1.4 Ratings and characteristics

The standards specify some or all of the following ratings or characteristics:

voltage
current
frequency
temperature rises or power dissipations of fuselinks
power acceptance of fuse-holders or bases
breaking capacity
time/current characteristics, gates or zones

current cut-off and I^2t characteristics
dimensions or size.

8.1.4.1   Voltage rating

The standardised or preferred rated voltages tend to be aligned with the system voltages which are commonly used throughout the world. Different philosophies are adopted, however, in relation to the rated and test voltages of the three categories of fuses.

High-voltage fuses, produced to IEC specifications, are assigned with rated voltages somewhat higher than the voltages of the systems in which they are to be used, for example, fuses rated at 12 kV are used in systems operating at 11 kV (line). These circuits are, of course, three-phase and in the event of a three-phase short-circuit one of the fuselinks will be the first to operate. When it has cleared, a line-to-line fault will exist between the two energised phases, which will thus have the same voltage at the fault position. As a result, and as shown in Figure 8.1, the voltage across the fuselink which has cleared will be V_{RF} which is 1·5 times the phase voltage (i.e. $1·5/\sqrt{3} \simeq 0·87$ of the line voltage) of the system. For this reason it is required that type tests be conducted at 87 per cent of the rated voltage of high-voltage fuselinks. A fuselink for use in an 11 kV system is therefore tested to ensure that it will operate satisfactorily in a circuit which provides a recovery voltage of 10·44 kV.

For low-voltage fuses, the IEC specifications require that fuselinks should be type tested at voltages 10 per cent higher than the assigned rated voltages and the latter may then correspond to the normal operating voltages of the networks in which they are to be used. These fuses are often used in single-phase circuits and in Britain fuselinks rated at 240 V and tested at 264 V are fitted in 240 V circuits. This arrangement and that adopted for high-voltage fuses ensures that the fuselinks will clear if the circuits happen to be operating slightly above their normal voltage levels when overcurrents occur. For 690 V applications, the test voltage is 5 per cent higher. The reason for this is that it stemmed from 660 V systems and that the voltage regulation

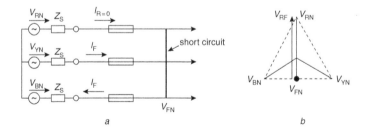

Figure 8.1   *Recovery voltage when first fuselink clears*
Where
a   circuit
b   phasor diagram
Z_s source impedor

is more controlled at this new voltage of 690 V. Furthermore, it aligns with associated switchgear standards such as *IEC 60947*.

Miniature fuselinks are usually rated at 250 V and are tested at this voltage. They are normally used on systems of lower voltage, e.g. 240, 220 or 208 V.

The different philosophies are so well entrenched that changes which will lead to alignment are not likely to be made in the near future.

Further confusion could arise because of the different national practices. The national standards of the UK and continental European countries are aligned with those of the IEC, but in the USA and associated territories all fuses are assigned rated voltages higher than those of the systems in which they are to be used.

The differences do not cause difficulties in countries in which users obtain their fuses from manufacturers with which they are familiar but, with the increase in competition and international trading, users should be aware that voltage ratings are not all determined on a single uniform basis.

8.1.4.2 Current rating

It is obviously undesirable to have a proliferation of current ratings and therefore particular or preferred ratings are specified in standards. There is a general tendency, particularly in the IEC standards, to follow the R10 series or, if necessary, the R 20 series may be used. These series simply divide a decade into 10 or 20 steps in geometric progression, the values being rounded off. As an example, using the R10 series, fuselinks with ratings from 10 to 100 A are produced for 10, 12, 16, 20, 25, 32, 40, 50, 63, 80 and 100 A. Some manufacturers and standards use the more accurate figures of 12·5 and 31·5 A for 12 and 32 A, respectively.

8.1.4.3 Frequency rating

To cover the power frequencies used in the various countries, the IEC specifications typically require that fuselinks should be able to operate satisfactorily in circuits with frequencies in the range 45–62 Hz. As stated earlier, fuselink behaviour is not significantly affected by the power-system frequency except at very high current levels where the I^2t values are somewhat dependent on it.

In addition to the above requirement, some standards specify that fuselinks should be capable of interrupting direct currents.

8.1.4.4 Temperature rises or power dissipation of fuselinks

Again there is not complete uniformity of practice in the specifications for the different categories of fuses.

For high-voltage fuselinks the requirement is that specified temperature rises must not be exceeded at the rated currents. Compliance with these requirements is checked during type tests in which a number of fuselinks are tested in standardised test rigs to ensure uniformity between the various manufacturers.

With low-voltage fuselinks, emphasis has been placed on power dissipation in recent years, as this enables their suitability for mounting in particular fuse-holders to be checked easily. Maximum power dissipations at rated-current levels are now specified in IEC standards for fuselinks produced to given dimensional standards. Again, compliance is checked during the type-testing procedure, a number of fuselinks being mounted in standardised test rigs to check that the maximum power dissipations do not produce excessive temperature rises.

Miniature fuselinks are treated differently, the maximum voltage drops across them at rated currents being specified. In this instance, temperature measurements need not be taken during the type testing.

8.1.4.5   Power acceptance of fuse-holders and fuse bases

The rated power acceptances of fuse-holders are established from type tests, the criterion being that the temperature rises of the component parts should not exceed permitted levels. In practice, fuselinks which have had the maximum permitted power dissipation when tested in the standardised test rigs, referred to above, are mounted in the fuse-holder under test and, provided that the temperature rises are not excessive, then the fuse-holder is assigned a power-acceptance rating equal to the maximum fuselink power-dissipation value.

8.1.4.6   Breaking capacity

During type tests, manufacturers prove that their individual designs are capable of interrupting particular high currents. These values are then designated the maximum breaking capacities of the designs and, as explained earlier, this does not imply that these fuselinks can interrupt all lower prospective currents down to the minimum fusing values. It is clearly not possible or required that there should be a single maximum breaking capacity for all fuselinks, but some standardisation has been introduced into IEC specifications.

Miniature fuselinks of high breaking capacity are required to be capable of interrupting alternating currents of 1500 A whilst those of low breaking capacity must be able to clear the greater of ten times their rated current or 35 A.

Low-voltage fuselinks for industrial use with rated voltages up to 500 V must be capable of interrupting currents of at least 50 kA and domestic fuselinks with ratings up to 415 V must be able to clear 20 kA.

Because of the varied usage of high-voltage fuses there are no maximum breaking capacities specified in the IEC standards.

8.1.4.7   Time/current characteristics

Specific characteristic curves have not been standardised and quoted because the constraints which might be imposed by doing so could stifle future development and prevent the introduction of new fuses with slightly different time/current characteristics. The modern trend is to specify a number of points which form gates through which the actual time/current characteristics of all manufacturers' fuselinks must pass if they are to comply with the appropriate standard. This method is illustrated in Figure 8.2 and applies to current IEC miniature fuse standards. In addition to gates,

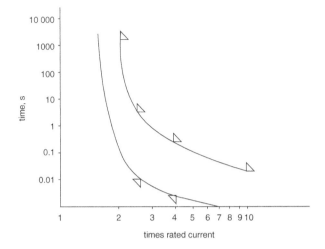

Figure 8.2   Standard time/current characteristic
Where
△   specified points

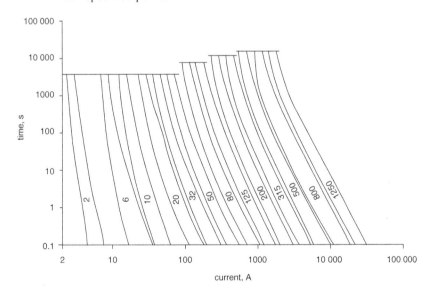

Figure 8.3   Time/current characteristic zones to BS88-Z

some of the IEC recommendations on low-voltage fuses specify zones within which all characteristics must lie. The zones were chosen so that all fuselinks of a given rating would operate in a shorter time at any current than any fuselink of at least 1·6 times the rating (two steps in the R10 series). Such time current zones specified in BS88-Z are shown in Figure 8.3.

8.1.4.8   Cut-off current and I^2t values

Manufacturers are required to quote these values for their various fuselinks. In the IEC low-voltage fuse standard pre-arcing I^2t limits are specified at 0.01 s for gG fuselinks and additional values are standardised for specific fuse systems. The miniature and high-voltage standards do not specify I^2t values.

8.1.5   Markings

It is clearly necessary that the markings on fuses and fuselinks should be permanent and, in addition, it is desirable, to prevent possible confusion and misunderstandings, that the information provided should be in a standardised form. The requirements associated with each of these matters are stated in the standards.

Obviously, there is a limit to the information that can be provided on fuses and fuselinks and it has been generally agreed that the following details are adequate:

current rating
voltage rating
name of supplier or manufacturer
manufacturer's type reference for the device
reference to the type of characteristics possessed by the device
standard with which the device complies.

The use of colour coding for fuses is not preferred by the IEC because of the possibility that colours may change quite appreciably over long periods of time and also because there are a surprisingly large number of people who are colour blind. Such colour coding is nevertheless used by individual manufacturers to enable easy identification of the fuses in their stores and also facilitate the initial installation into equipment.

IEC 60127-1 (miniature fuses) describes an optional colour banding scheme in Appendix A, but this is combined with mandatory alpha-numeric coding in *IEC 60127-2* for cartridge fuselinks. This coding consists of the following sequential markings:

- the symbol denoting the relevant pre-arcing time/current characteristic – FF, F, M, T or TT (see Section 6.1.1)
- the rated current
- the symbol denoting the rated breaking capacity – H, L or E (see Section 6.1.1)
- the rated voltage.

Examples of markings are:

T	3	1	5	L	2	5	0	V
		F	4	H	2	5	0	V
T	3	1	5	E	2	5	0	V

Colour coding is specified for the fuse indicator for 'D' type fuses to *IEC 60269-3-1 Section I* which assists in the non-interchangeability of current ratings.

NH fuselinks to *IEC 60269-2-1* Section I have specified colours of marking as described in the following table:

Characteristics	gG		aM	
Colour marking	Black		Green	
Kind of print	Strip with inverse print	Normal print	Strip with inverse print	Normal print
Voltage (V)				
400[1]	X		X	
500		X		X
690	X		X	

[1] For 400 V gG, a blue colour is also permitted.

8.1.6   Type tests

Fuselinks, unlike most other equipment, cannot be subjected to extensive routine proving tests at the end of the production process because if they are operated they cannot be used again. The behaviour of individual designs must therefore be determined by very rigorous type tests and then the subsequent component parts must be produced to within very close limits of those used in the type-tested fuselinks. In addition, inspection and quality assurance systems must be employed to ensure that the volume-production output corresponds closely with the initial devices. These systems are discussed in detail in Chapter 9 and only the type-test procedures are examined in this chapter.

The main part of each standard is usually devoted to the type tests. In service, the conditions encountered by fuselinks may be very variable and, of course, their performances may be affected thereby. To ensure uniformity between manufacturers, type tests must be done in specified and standardised conditions. The tests are conducted in laboratories and, because of the control which is available, the limits set on parameters such as ambient temperature are much smaller than those specified in the standard conditions of service.

Because the performances of fuselinks may be significantly affected by factors such as the impedances of the test circuits, the size and disposition of attachments (including the cables) and the proximity of supports or enclosures, the standards specify the test arrangements in great detail and this is particularly so for the time/current and short-circuit-breaking-capacity tests.

The actual type tests which are undertaken depend on the fuselink being examined, but in general the following checks and tests are done on a number of pre-production fuselinks.

Fuselinks in a given size and construction may cover a series of current ratings. A reduction in type tests to the standards is permitted, if specified rules apply, these rules apply where such fuses are part of a homogeneous series of fuselinks. This is defined as 'a series of fuselinks within a given size, differing from each other only in such characteristics that for a given test, the testing of one or a reduced number of particular fuselinks of that series may be taken as representative for all the fuselinks of the series itself'. In particular, the elements must consist of identical materials, they shall have the same form, e.g. with identical tools but may be from material of different thickness.

If the specified conditions in the product standards are met, then it is generally only necessary to undertake a full set of type tests on the maximum current rating and reduced testing on the smaller current rating.

8.1.6.1   Construction and dimensions

Each fuselink which is to be used during type testing must be carefully examined during manufacture to ensure that there is nothing abnormal in its construction and the dimensions of the component parts are measured accurately to see that they are within close tolerances of the values to be used in the subsequent volume production.

8.1.6.2   Electrical resistance

This value is measured for each fuselink to be type-tested. It must be done at an ambient temperature, typically in the range 20–25°C.

8.1.6.3   Power dissipation

Normally, low-voltage fuselinks must be tested in a standard rig, as stated earlier. The power dissipated by each fuselink must be measured at rated current and the temperature rises of contacts or terminals be measured by thermocouples.

Similar tests are done on high-voltage fuselinks but the power input need not be measured and when miniature fuselinks are tested only the voltage drop at rated current need be recorded.

8.1.6.4   Power acceptance of fuse-holders

This test, which is only applied to low-voltage fuses, is performed by taking the fuselink found to have the highest power dissipation in the above test and inserting it into a fuse-holder. Rated current is then passed through it again and the temperature rises at various points are measured by thermocouples.

8.1.6.5   Insulation levels

Extensive tests must be done on high-voltage fuses because of the vulnerable positions in which they may be used. They are subjected to impulse voltages ($1/50\,\mu s$ waves in the UK) between parts which will, in service, be live and earthed, respectively. They are also subjected to power-frequency over-voltages under both wet and dry conditions.

Low-voltage fuselinks in their fuseholders are mounted on a metal panel and a power-frequency supply of $2\cdot5\,kV$ is connected for one minute between the panel and normally live fuse parts.

8.1.6.6 Conventional fusing currents

The terms 'conventional fusing current' and 'conventional non-fusing current' have been introduced into IEC low-voltage-fuse specifications to replace the term 'minimum-fusing current', which has been in use for many years. The latter current is strictly the one which will cause a fuselink to operate in an infinite time and therefore its determination is impractical. *BS 88* defines minimum-fusing current as that current which causes operation in four hours, but even the determination of this value is very time consuming. To simplify the situation, the IEC specifications require that all of a number of fuselinks mounted in a standard type-test rig should operate in less than the conventional time when they are carrying the conventional fusing current, and not operate in the conventional time when carrying the conventional non-fusing current. The conventional time, which may vary between one and four hours, depending on the current rating, is specified in the standards.

The concept of conventional currents has been used on the continent of Europe for several years and it is likely that the present British usage of the term 'minimum-fusing current' will be phased out. It has nevertheless been used in the earlier chapters of this book because its significance is immediately obvious.

8.1.6.7 Breaking capacity

Because fuses are usually the ultimate back-up protection in the circuits in which they are included, they must be capable of operating under the most onerous conditions which may arise. For this reason the tests at maximum breaking capacity are done under specified conditions in low-power-factor (typically less than 0·2 for low-voltage fuses) single-phase, inductive circuits arranged as shown in Figure 8.4. Equipment is included to enable the test circuits to be closed at any desired point in the voltage cycle so that conditions of varying severity may be produced. The fuselinks are mounted in standard rigs, an example for miniature links is shown in Figure 8.5. The tests determine not only breaking capacity but parameters such as I^2t let-through, arc voltage and cut-off current and, when the latter is being found, the circuit is switched so that arcing commences just prior to an instant when the system voltage is at its peak value. To cater for three-phase applications the appropriate source voltage is used in the test circuit, that is 87 per cent of the rated voltage for high-voltage fuselinks or 10 per cent more than the rated line voltage when low-voltage cartridge fuselinks are

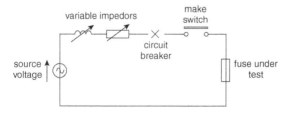

Figure 8.4   Test circuit

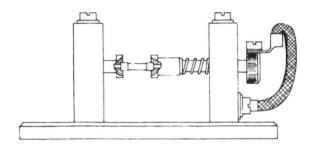

Figure 8.5   Standard test rig for miniature fuselinks

being tested. In all breaking-capacity tests the full recovery voltage is maintained for at least 30 s and for high-voltage fuses it is maintained for 60 s.

In addition to the above tests, the specifications require manufacturers to do type tests at lower current levels, and for these, the power factors of the test circuits are generally higher than those used at the maximum currents. This represents the situation which tends to arise in service, proportionally more resistance being present at the lower fault levels. This is also recognised in the IEC specifications which allow power factors of 0·3–0·5 for low-voltage fuselinks. Minimum breaking capacities are not usually quoted by manufacturers for miniature fuselinks and for the majority of low-voltage, general-purpose fuses, but a number of type tests are done at different currents to ensure that they will operate satisfactorily at somewhat arbitrarily chosen multiples of the rated currents. The minimum breaking capacity of high-voltage cartridge fuselinks is not specified in the IEC documents, but it is nevertheless an important parameter which is usually quoted by the individual manufacturers after conducting tests in the standard manner.

To conduct the maximum-breaking-capacity tests, short-circuit testing stations of high MVA output are required to provide the high currents needed at the required voltages. Such stations are usually employed also for the testing of switchgear. Both these uses only require the high outputs for short periods and so in the interests of economy the stations are short-time rated. As a result, problems can arise when minimum-breaking-capacity tests are attempted because the fuselink operating times may then be of the order of one hour and during such periods the testing-station equipment could overheat. Recognising this situation, the IEC specifications permit a two-part test method, in which the fuselinks are supplied initially with the desired current from a separate very low-voltage source and when the fuselink element is almost at the melting temperature, the circuit is switched to obtain the current from the short-circuit alternator which is excited to provide the required recovery voltage.

Fuselinks which exhibit current limitation are subjected to a further important breaking capacity test which simulates maximum arc-energy conditions in the current-limiting region. This is sometimes referred to as the critical current because it can often be more onerous than the current corresponding to the maximum breaking capacity. It is also referred to as Test Duty 2 or I_2 in the IEC specifications. The conditions which cause the arc energy to be a maximum have not been discussed earlier in this text and because of the importance of this particular matter they are considered in some detail below.

When a fuselink is clearing a fault, there is energy stored in the inductive elements in the circuit at the time when arcing commences in the fuse. This energy, given by $\frac{1}{2}Li_0^2$, in which i_0 is the current at the start of arcing, has to be dissipated partly in the circuit resistance but mainly by the fuse during the arcing process.

In a faulted circuit of low-power factor, the prospective current is almost inversely proportional to the circuit inductance and therefore the stored energy at any instant would, in the absence of current limiting, be proportional to the prospective current (I). However, under current-limiting conditions with high prospective currents, the cut-off current tends to be approximately proportional to the cube root of the prospective current and therefore the stored energy at the beginning of arcing is given by

$$\tfrac{1}{2}\,Li_0^2 \propto Li^{2/3} \propto I^{-1/3}$$

This approximate relationship shows that the stored energy decreases with increase of prospective current over the range of currents where current limitation occurs. At lower currents, however, it is proportional to the prospective current, as stated earlier. The variation is thus of the form shown in Figure 8.6, from which it can be seen that there is a current at which the stored energy is greatest. Of course, the actual arc energy is greater than the stored inductive energy at arc initiation because additional energy is transferred to the fuselink, from the source, during the arcing period. This does not, however, so alter the situation that the maximum energy to be dissipated by an arc occurs at a prospective current below that corresponding to the maximum breaking capacity.

In practice, experience has shown that the maximum-energy condition usually occurs at a prospective current of about three times that needed to give a pre-arcing time of one half-cycle.

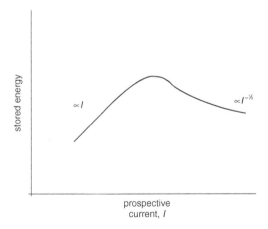

Figure 8.6   Variation of stored energy with prospective current

8.1.6.8 Time/current characteristics

Compliance with the specified gates or zones is checked by the manufacturers during the breaking-capacity tests by operating fuselinks at the necessary current levels and noting the operating times.

8.1.6.9 Overload withstand capability

Some standards require that tests be done, using the standard rigs employed for checking power dissipation at rated current, to check the ability of fuselinks to carry specified overload currents. Such tests apply to miniature, low-voltage and high-voltage motor fuselinks and, for example, for miniature fuses the test consists of 100 cycles, each of 1·2 times rated current for 1 h followed by 15 min of zero current, and then finally an hour at 1·5 times rated current. This test must be done on three fuselinks and at the end their voltage drop must not have changed by more than 10 per cent. The requirement for industrial and domestic low-voltage fuselinks is that they should be subjected 50 times to a cycle of 5 s at 80 per cent of the minimum operating current followed by zero current for 20 per cent of the conventional time.

All the above type tests do not have to be done on every current rating of each type of fuselink because, for example, certain tests may not relate to or require the passage of current. The standards therefore specify, in detail, the number of tests to be done and the current ratings of the fuselinks which are to be tested. To limit the amount of testing and to cut costs, by minimising the tooling required, it is common practice to design several current ratings of fuselinks with the same element geometry by using either different element thicknesses or different numbers connected in parallel. If the variations conform to certain specified rules, a reduced number of type tests may be made on such a family or series of ratings. These groups are called homogeneous series and they allow the number of breaking-capacity tests done on them to be reduced significantly.

The allowable limiting values of specific parameters are either specified within the test clauses or, alternatively, included in further sections of the standards. It is not practical to refer to all the limits in this work, and indeed it is unnecessary because they can be obtained by referring to the appropriate standards. A number of values have been included in the preceding sections, however, to give indications of the magnitudes of the test-current levels and the durations of test programmes.

8.1.7 Dimensions

Although there is a desire to have fuse dimensions standardised on an international basis, it has not so far been possible to achieve this, because of the inherent historical variations and designs which now exist and the need to continue to produce them in large numbers for replacement purposes. Manufacturers are naturally reluctant to increase the numbers of designs which they have to produce. As a result, the IEC has only been able to list the commonly used dimensions on Data Sheets, which are only for information and not in any way mandatory.

8.1.8 Application guides

The main task of IEC standards is to formulate test requirements to cover a wide variety of needs. Having achieved this, the fuse standards are now mostly finalised rather than subject to radical new revision. However, later attention has been focused on 'Application Guides' and information to assist the user. In addition, within the IEC Fuse Standards, particularly in the Annexes, there is a lot of useful information relating the standardised test requirements to practical applications.

8.1.8.1 Low-voltage fuses

In 2003, the *IEC TR61818* was published. This technical report gives additional information on fuses as well as important facts which are sometimes difficult to locate in the standards. The guide also makes reference to other associated IEC standards and publications.

The important benefits of current-limiting cartridge fuselinks are outlined and these have already been covered in Section 7.1.

A section on fuse combination units describes the associated products standardised in *IEC 60947-3* and the following table illustrates the functions, definition references and symbols of various types of fuse combination units. Please note that the modern terminology replaces 'isolator' by disconnector.

Function		
Connecting and disconnecting	Isolating	Connecting, disconnecting and isolating
switch 2.1 of IEC 60947-3	disconnector 2.2 of IEC 60947-3	switch-disconnector 2.3 of IEC 60947-3
fuse-combination unit, 2.4 of IEC 60947-3		
switch-fuse 2.5 of IEC 60947-3	disconnector-fuse 2.7 of IEC 60947-3	switch-disconnector-fuse 2.9 of IEC 60947-3
fuse-switch-fuse 2.6 of IEC 60947-3	fuse-disconnector 2.8 of IEC 60947-3	fuse-switch-disconnector 2.10 of IEC 60947-3

Two different types of fuse combination units exist:

(*a*) Switch-fuses and switch-disconnector-fuses with switches connected in series with the fuselinks and are devices with dependent manual operation (snap action).

(*b*)   Fuse-disconnectors and fuse-switch-disconnectors which use the fuselink itself to form the moving part and are of dependent manual operation (operator dependent).

Since most of the fuse combination units with the fuse as an integral unit are designed as fuse-switch-disconnectors, or switch-disconnector-fuses, they may be used for:

- load switching
- isolation
- short-circuit protection.

Types of low-voltage fuselinks are designated by the following letters:

- The first letter describes the breaking ranges:
  a – partial range: all currents between the lowest current indicated on its operating time current characteristic and its rated breaking capacity.
  g – full range: all currents which cause the melting of the fuse element up to its rated breaking capacity.
- The second letter describes the application (characteristics or utilisation category).

The various types are 'buried' within the standards and below is a table updating the information given in *IEC TR61818*:

Type	Application (characteristic)	Breaking range
gG	General application, mainly for conductor protection	Full range
gM	Motor circuit protection	Full range
gN	North American general application for conductor protection	Full range
gD	North American general application time delay	Full range
gR, gS	Semiconductor protection	Full range
gU	Wedge tightening fuse for utilities	Full range
gL, gF, gI, gII	Former type of fuse for conductor protection (replaced by gG type)	Full range
aM	Short-circuit protection of motor circuits	Partial range (back-up)
aR	Semiconductor protection	Partial range (back-up)

The technical report advises for popular systems, the 'maximum operational voltage of the fuselink'. Typical maximum operational voltages, which equate to the test voltage, are given in the following table.

This clarifies the 690 V situation as outlined in Section 8.1.4.1. In addition, it shows that the maximum operational voltage for gN and gD fuses is the same as the rated voltage – 600 V. The gN and gD are fuse systems used in North America where commonly used nominal system voltages are 460 or 480 V, so in practice, there is

Type	Rated voltage (V)	Maximum operational voltage (V)
gG, gM, aR, aM, gR, gS	230	253
	400	440
	500	550
	690	725
gN, gD	600	600

a safety factor between the nominal system voltage and the maximum operational voltage. Furthermore, if the gN and gD fuselinks were used in systems outside of North America the nominal system voltages are typically 400 V, so again there is a significant safety factor.

8.1.8.2   High-voltage fuses

Current-limiting fuses

Application guidance is given in *Clause 8 of IEC 60282-1* and presents suggestions on application, operation and maintenance as an aid in obtaining satisfactory performance with high-voltage, current-limiting fuses.

It cautions that high-voltage fuses should be handled with at least the same degree of care as any other precision made item of equipment (such as a relay). In addition, if during internal installation, fuses are subjected to severe mechanical stresses, then special tests may be agreed by the user and manufacturer of the fuses and the switchgear. For switch fuse combinations, see *IEC 60420*.

A section includes the selection of the rated current of the fuselink, although as was seen in Sections 7.4, 7.5 and 7.7, the chosen rated current for the fuselink in applications are generally dictated by transient phenomena in the circuit related to switching, such equipment as transformers, motors or capacitors.

However, fuselinks are inherently and invariably mounted in more adverse thermal conditions than the standardised test arrangements, for example, in enclosed fuse 'pods'. A typical de-rating can be 25 per cent of the rated current, so in such applications the service current carrying capability of the fuselink has to be assessed.

The rated voltage should be selected with regard to:

* Three-phase solidly earthed neutral systems – the highest line-to-line voltage.
* Single-phase systems – 115 per cent of the highest single-phase system voltage.
* Three-phase isolated neutral systems – testing is necessary at higher than the 0·87 times the voltage rating of the fuse, as specified in the standard.
* Capacitive currents – in phase-to-earth faults.

With regard to operation, special care should be taken to see that the fuselink is securely locked in the service position. It is also advisable to remove and insert fuses when in an off-load de-energised condition.

Clause 8.3.4 of *IEC 60282-1* covers selection according to class: back-up, general purpose or full range and minimum breaking current.

The standard has a number of 'normative' and 'informative' annexes. The 'informative' annexes are:

Annex B: Reasons which led to the choice of transient recovery voltage (TRV) values for test duties 1, 2 and 3.

Annex C: Preferred arrangements for temperature rise tests of oil-tight fuselinks in switchgear.

Annex D: Types and dimensions of current-limiting fuselinks specified in existing national standards.

Annex E: Determination of de-rating when the temperature surrounding the fuse exceeds 40°C. This gives a worked example that is worthy of examination and could well apply to low-voltage installations.

Expulsion fuses

Application guides are given in Clause 11 of *IEC 60282-2* covering expulsion fuses and present suggestions on the application, operation and maintenance as an aid in obtaining satisfactory performance with expulsion and similar fuses.

The reader is reminded that drop-out fuse carriers that remain in the open position for prolonged periods of time, may accumulate water and pollution in their internal parts, which may result in the degradation of their operational properties.

Fuses should be mounted in the position specified by the manufacturer and precautions taken to allow for the high noise level and emission of hot gases. Like current-limiting fuses, the current ratings of expulsion fuses are generally dictated by transient current phenomena in the circuits they are protecting.

There are three classes of fuses based on the protection of the associated transformers, capacitor banks and feeder circuits.

Class A: Remotely placed from major substations.
Class B: Close proximity to major substations.
Class C: Close proximity to major substations without parallel connected loads.

Some national standards include additional requirements, including special applications, these include:

- Spark production tests
- Robustness
- Forces required to open and close drop-out fuses
- Current surge withstand tests.

The standard includes three 'informative' annexes.

Annex A: Reasons for the selection of breaking test values.
Annex B: Typical fuselink dimensions.
Annex C: Operating rods.

8.1.8.3   Miniature fuses

A User Guide, *IEC 60127-10* was published in 2002. The object is to introduce the user to the important properties of miniature fuselinks and fuse holders and to give some guidance in applying them.

The properties of miniature fuses are highlighted:

- Protecting upstream, isolating downstream and diagnosing fault location.
- Wide range of fuselinks and holders.
- Lower cost and small dimensions.
- Wide range of tamper-proof and reproducible characteristics.
- Discrimination (selectivity).
- Safe and reliable.

Unlike power systems, it is often difficult to calculate the maximum potential fault current of a circuit application. Often it is an assumed theoretical value, assigned by a safety agency. In some cases, the suitability of a fuse's breaking capacity is determined by testing the fuse in the end product, under short-circuit conditions.

The Guide has tables, standard sheets of standardised fuselinks in *IEC 60127-2, 3 and 4*, giving salient properties for ease of reference.

Some good general guidance is provided on fuse selection, which is usually dictated by three basic categories of criteria.

(*a*)   Electrical requirements of the application, including I^2t and DC requirements.
(*b*)   Conformance to published safety standards.
(*c*)   Mechanical properties/physical size.

The User Guide includes a section on direct current (DC) and typical DC applications include:

- Batteries/accumulators which are comparatively low voltage (less than 50 V) but with potentially high fault currents.
- Telecommunications or power supplies up to 125 V where the fault current is within the AC breaking capacity of the fuselink.
- DC voltages above 125 V where additional testing may be necessary, particularly for breaking capacity.

The time constants of such currents are usually less than 2 ms for battery circuits and up to about 4 ms for other inductive circuits that can typically be protected by miniature fuses.

Attention is drawn to the selection of the thermal rating of the fuse-holder. It is based on the maximum power acceptance of the fuse-holder, taking into account the local thermal conditions. The maximum *sustained* dissipation of the fuselink shall be less than, or equal to, the admissible power dissipation of the fuse-holder.

8.2   IEC fuse standards

8.2.1   Low-voltage fuses

These are covered by the *IEC 60269 series. Part 1* covers general and common requirements for all types of fuses. *Parts 2, 3 and 4* then cover requirements for industrial, domestic and semiconductor protection, respectively.

A list of the IEC fuse standards is included below:

60269-1	General requirements.
60269-2	Supplementary requirements for fuse for use by authorised persons.
60269-2-1	Examples of standardised fuses.
60269-3	Supplementary requirements for fuses for use by unskilled persons.
60269-3-1	Examples of standardised fuses.
60269-4	Supplementary requirements for fuselinks for the protection of semiconductor devices.
60269-4-1	Examples of standardised fuses.

Dimensional and further technical requirements, particular to a fuse system, are covered in *Parts 2-1, 3-1 and 4-1*. For example in *Part 2-1*:

Section I	Fuses with fuselinks with blade contacts (NH fuse system).
Section II	Fuses with fuselinks for bolted connections (BS bolted fuse system).
Section III	Fuses with fuselinks having cylindrical contact caps (NF, French, cylindrical fuse system).
Section IV	Fuses with fuselinks with offset blade contacts (BS clip-in fuse system).
Section V	Fuses with fuselinks having 'gD' and 'gN' characteristics. (Class J and class L time delay and non-time delay fuse types.)
Section VI	'gU' fuselinks with wedge tightening contacts.

Other fuse standards are:

60146-6	Application Guide for the protection of semiconductor converters against over-current by fuses.
61459	Co-ordination between fuses and contactors/motor starters – Application Guide.
61818	Application Guide for low-voltage fuses.

8.2.2   High-voltage fuses

High-voltage fuse standards are covered by the *60282 series*

60282-1   Current-limiting fuses.
60282-2   Expulsion fuses.

In addition, there are the following standards relating to specific applications:

60549   External protection of shunt capacitors.
60644   Motor circuit application.
60787   Transformer circuit applications.

8.2.3   Miniature fuses

Miniature fuse standards are covered by the *60127 series*:

60127-1   General requirements.
60127-2   Cartridge fuselinks.
60127-3   Subminiature fuselinks.
60127-4   Universal modular fuselinks (UMFs).
60127-5   Guide for quality assessment.
60127-6   Fuse-holders.
60127-10   User Guide.

8.2.4   Temperature rise

IEC 60943 is a technical report covering guidance concerning the permissible temperature rise for parts of electrical equipment, in particular for terminals and was prepared by the Fuse Committee. It is more of a 'classical' reference document covering the temperature rise in electrical assemblies, generally and not specifically to fuses. The report is split into two sections:

- Theory
- Applications.

And these are supported by additional information in the annexes.
   The report is intended to supply:

- general data on a structure of electric contacts and the calculation of their ohmic resistance.
- The basic ageing mechanisms of contacts.
- The calculation of the temperature rise of contacts and connection terminals.
- The maximum 'permissible' temperature and temperature rise for various components, in particular the contacts, the connection terminals and the conductors connected to them.
- The general procedure to be followed by product committees for specifying the permissible temperature and temperature rise.

8.3   European standards

Countries belonging to the European Economic Area (EEA) – European Union (EU) and European Free Trade Association (EFTA), are duty bound to have common European standards wherever possible, based on IEC standards. In addition, there are 12 affiliates from Central and Eastern Europe.

The standards making body for Europe is CENELEC – European Committee for Electrotechnical Standardisation. European standards within the EEA are given the letters 'EN', standing for Euro Norm. CENELEC does not publish ENs, but issues a cover document stating they comply with the appropriate IEC standard. The EN thus has the designation of EN followed by the IEC reference, for example:

EN 60127-1
EN 60269-3
EN 60282-1

They can be obtained from National Committees and the reference is preceded by the national reference, e.g. *BSEN, DINEN.*

Where there are national deviations, then a Harmonised Document, HD, is produced where the deviations from the standard are itemised.

Only two of the IEC standards described in Section 8.2 are not ENs. These are *IEC 60269-2-1 and 60269-3-1.* In these cases, National Committees may select one or more systems (sections) for their national standard.

The references to these standards are:

HD 630.2.1 (IEC 60269-2-1)
HD 630.3.1 (IEC 60269-3-1)

8.3.1   British standards

Major British standards align with EN standards and in turn, IEC standards. This applies to the *60269 (LV), 60282 (HV)* and *60127 (Miniature)* standards. However, in the low-voltage field, the old British standard numbers are retained, including *BS88 (Industrial) and BS1362 (Plug Top).* This is due to their references in the National Wiring Regulations *BS7671 (IEC 60364)* and in the case of *BS1362* for Plug and Socket legislation. In addition, there is the issue of familiarity. However, in time the British standard references will be phased out as indeed has occurred in the miniature fuse field to the IEC standards. To assist in this transition, the British Fuse standards have three references for example:

BSEN 60269-1
IEC 60269-1
BS 88-1

There are some British fuse standards that are unique to the UK and it is permissible to continue with these standards. A list of BS fuse standards are as follows:

BS	EN (HD)	Description
88-1	*60269-1*	General requirements
88-2.1	*60269-2*	Supplementary requirements for fuses for use by authorised persons
88-2.2	*HD 630.2.1 Section II (IEC 60269-2-1)*	Additional requirements for fuses with fuselinks for bolted connection
88-3.1	*60269-3*	Supplementary requirements for fuses for use by unskilled persons
88-4	*60269-4*	Supplementary requirements for fuselinks for the protection of semiconductor devices
88-5	*HD 630.2.1 Section VI (IEC 60269-2-1)*	Supplementary requirements for fuselinks for use in AC electricity supply networks
88-6	*HD 630.2.1 Section IV (IEC 60269-2-1)*	Supplementary requirements for fuses of compact dimensions for use in 240/415 V AC industrial and commercial electrical installations
646		Cartridge fuselinks (rated up to 5 A) for AC and DC services
714		Cartridge fuselinks for use in railway signalling circuits
1361	*HD 630.3.1 Section IIB (IEC 60269-3-1)*	Cartridge fuses for AC circuits in domestic and similar premises
1362	*HD630-3-1 Section IV (IEC 60269-3-1)*	General purpose fuselinks for domestic and similar purposes, primarily for use in plugs
3036		Semi-enclosed electric fuses (ratings up to 100 A and 240 V to earth)
7656		Low-voltage pole-mounted fuses (cut-outs) 315 A rating
7657		Fuses (cut-outs) ancillary terminal blocks and interconnecting units up to 100 A rating for power supplies to buildings

8.3.2   Other national standards

A similar position to the UK applies in other European countries, for example, in Germany the NH fuselinks are covered by

DIN EN	*60269-1*
IEC	*60269-1*
VDE	*0636-10*

8.4   North American standards

The USA, Canada and Mexico are members of the IEC, with USA and Canada being active participants of the IEC fuse committees. North American practices are somewhat different to those of Europe, which in turn strongly influence most of the rest of the world. The USA consumes over 25 per cent of the world's production of electricity and therefore an appreciation of its practices in relation to fuses is appropriate. The USA National Standards Committee is ANSI, American National Standards Institution and in Canada it is CSA, Canadian Standards Association. Canada's practices tend to follow those of the USA but historically there are British and French influences.

8.4.1   Low-voltage and miniature fuses

Although the National Committee in the USA is ANSI, the most influential body for fuse standards is UL, Underwriter Laboratories. UL produces standards and is a leading approvals body. The technical differences in the USA and IEC requirements for fuses stem from the Wiring Regulations which in the USA is the NEC, National Electric Code and further progress in alignment of the UL and IEC fuse standards inherently has to await progress on the alignment of the Wiring Regulations. A list of UL fuse standards follows, which combines the low-voltage and miniature categories.

248-1	*Part 1:*	General requirements
248-2	*Part 2:*	Class C fuses
248-3	*Part 3:*	Class CA and CB fuses
248-4	*Part 4:*	Class CC fuses
248-5	*Part 5:*	Class G fuses
248-6	*Part 6:*	Class H non-renewable fuses
248-7	*Part 7:*	Class H renewable fuses
248-8	*Part 8:*	Class J fuses
248-9	*Part 9:*	Class K fuses
248-10	*Part 10:*	Class L fuses
248-11	*Part 11:*	Plug fuses
248-12	*Part 12:*	Class R fuses

248-13	*Part 13:*	Semiconductor fuses
248-14	*Part 14:*	Supplemental fuses
248-15	*Part 15:*	Class T fuses
248-16	*Part 16:*	Test limiters

Miniature fuses are covered by *248-14* Supplemental fuses.

8.4.2   *High-voltage fuses*

The most influential body for high-voltage fuses in the USA is the IEEE and the fuse committee is a subcommittee of the switchgear committee. The IEEE publishes standards on fuses which are covered in the *ANSI C37 series*.

8.5   Approvals procedure

The various standards referred to in the previous sections specify limiting values and test procedures, but in the interests of all concerned it is necessary that users are satisfied that the required type tests have been done on their fuses, that the results are acceptable and that the necessary standards are maintained during volume production.

Rather than arrange for manufacturers and all individual users jointly to witness type tests and inspect production fuses, and so that the number of approval procedures can be minimised, it is the custom and practice for standardised miniature and low-voltage fuses used in domestic and similar applications to be independently tested by approvals bodies, such as ASTA Certification Services in the UK. Similar bodies exist in Germany and North America, namely:

- Verband Deutscher Electrotechniker (VDE) in Germany
- Underwriters Laboratories Inc (UL) in USA
- Canadian Standards Association (CSA) in Canada.

These approval bodies conduct independent type tests, as well as market surveillance and quality assurance checks, and also, possibly, follow-up tests. They allow their initials to be used as approval marks on products with which they are satisfied.

In order that type tests are not repeated by the various bodies and in particular that they may be tested to the same IEC standard, ever increasing use is made of the IECEE (World-wide system for Conformity Testing and Certification of Electrical Equipment) CB Scheme. This is a well-established procedure governed by specified rules.

These approval procedures are generally spreading to the low-voltage industrial fuses and now in Germany and North America many fuses in this category must be approved by VDE, UL, or CSA and must bear their mark.

In the UK, it is the custom and practice to have the short-circuit-breaking-capacity tests on high- and low-voltage industrial fuses undertaken and/or witnessed by an independent body. A report which includes a certificate of short-circuit rating, of the

Rating Certificate No. 8839

ASTA The Association of Short-Circuit Testing Authorities

(Incorporated in the year 1938)   23/24 Market Place, Rugby CV21 3DU

Laboratory Ref. No. 4109-01 and
B/1466.

Certificate of Short-Circuit Rating

of   general purpose, non indicating cartridge fuse-links.

Short-Circuit Type Tested in accordance with   BS 88:Part 2:1975.

Rated Voltage   660 volts ac   Max. Rated Normal Current   630 and 100   amperes

Maker   Brush Fusegear Limited, Burton-on-the-Wolds, Loughborough, Leics.

Tested for   Brush Fusegear Limited.

Designation   R11   Serial No.   —

The apparatus, constructed in accordance with the description, drawings and photographs sealed and attached hereto, has been subjected by

Falcon Short-Circuit Testing Laboratory Limited.
British Short-Circuit Testing Station.

to a complete series of proving tests of its short-circuit rating which has been made, subject to any observations in the record, in accordance with the appropriate clauses of the Specification(s).

The results are shown in the RECORD OF PROVING TESTS and by the oscillograms sealed and attached hereto. The values obtained and the general performance are considered to justify the Short-Circuit Rating assigned by the manufacturer, as stated below.

Breaking-capacity at   0.660   kV. ac   Making-capacity   —   kA. peak at   —   kV.

Symmetrical   80   kA.   Short-time current capacity   —   kA.

(Equivalent to   —   MVA.)   for   —   seconds.

Asymmetrical   —   kA.   BS Reference : C2

Duty   Break.

This Certificate applies only to the short circuit performance of apparatus made to the same specification and having the same essential details as the apparatus tested.

The documents under seal forming part of this Certificate are:

(1) Record of Proving Tests: Sheets Nos.   1 to 10.

(2) Oscillograms Nos.   257203, 257206 − 257212, 257214 − 257216, 1/20849, 1/20851,
C5779713.   1/20853.

(3) Drawings Nos.

(4) Diagrams Nos.   4109/1 and 41RC586.

Photographs Nos.   15553, 15554, 15702, 15703, B2496, B2497, B2498.

*   The above fuse-links represent the maximum and minimum ratings of a homogeneous series. Fuse-links having intermediate ratings of 560, 500 and 450 amperes have been examined and comply with Clause 8.1.5.3 as part of this series.

ASTA Observer

Secretary

Date   30ᵈ October 1980

The conditions under which this Certificate may be reproduced are governed by Clause 14 of ASTA Publication No. 'Conditions for Test Work'.

Figure 8.7   ASTA certificate

form shown in Figure 8.7, is issued to show the tests undertaken and the products covered. The test results are quoted and a detailed drawing of the fuses that were tested is included.

Most European countries have a central short-circuit-testing station or organisation controlling several short-circuit-testing stations. In the UK, the practice to mark low-voltage fuses, ASTA 20 CERTIFIED or ASTA 20 CERT, if an ASTA certificate is available covering the design and the endorsement is authorised by ASTA, including initial and regular inspection.

For low-voltage fuses in the USA which are not standardised, for example, semiconductor fuses, it is possible to obtain UL component recognition where test results or specifically agreed tests have been witnessed or approved by the UL.

Chapter 9

Manufacture, quality assurance and inspection

The standards set for the construction of prototype fuses in both materials and manufacturing processes become the standards which must be maintained when bulk production commences. This is essential to ensure that the production fuses will have the performance characteristics indicated by the type tests and that they will conform with the appropriate specifications. This involves the preparation of detailed specifications and procedures for all stages of manufacture from the purchase of materials and components, through production processes to final inspection and testing. In parallel with this, inspection routines and procedures are set up to monitor current standards and achievements continually in order to provide assurance that standards are being met and that a continuous feedback of corrective action is applied to the earlier stages in the production process to maintain the most economical course towards achieving the set objectives.

The procedures adopted by various manufacturers obviously vary in detail but comply with the requirements of *ISO 9001: 2000*. Typical quality assurance/quality control practices employed are outlined in the following sections.

9.1 Quality assurance

Before describing the practices in detail the objects of the quality-control and inspection functions are given and the Acceptable Quality Level (AQL) for the products outlined.

Quality is best defined as conformance to an agreed specification. Quality control is the means whereby quality is assured through the maintenance of standards of quality through all the functions which can affect the quality of the finished product – through design, development, production engineering, production planning, actual production, packaging, inspection and even storage.

Confusion often exists over the terms 'quality control' and 'inspection'. Not only are these not synonymous expressions; they are in a sense diametric opposites. It is the object of inspection to detect faulty items. The object of quality control is to ensure that faulty – and wasteful – production does not occur. A perfect quality-control system would require no inspection.

However, perfection is impossible and in an actual production unit inspection is an important tool of quality control. Its use as a tool of quality necessitates its findings being recorded and rapidly assessable in order to effect immediately the improvements or corrections it shows as necessary.

Modern trends and introduction of cell manufacturing techniques have allowed much of the traditional inspection to be replaced by quality-assurance auditing with line operatives being given responsibility for their work.

All designs and manufacturing techniques should aim towards zero defects and the use of preventive tools such as failure-mode-and-effects analysis in both design and manufacturing is key.

Most companies, however, still use the well-proven sampling and AQLs.

When the AQL of the finished product has been determined, AQLs must be defined for all functions on a basis necessary to ensure the AQL of the product. For quantity production of comparatively inexpensive products it may be decided that a controlled wastage may have to be accepted rather than the expense of the refinements necessary to deliver products of the ultimately planned AQL to final inspection. This aspect makes final inspection very important, since it has to find the faulty components which make up acceptable wastage.

For example, if it is required to deliver to customers at an AQL of 0·4, i.e. with a possibility of four faulty products in 1000, it may be decided to deliver to final inspection at an AQL of 2·0 and suffer the loss of up to 2 per cent of products made. Because of the number of components required to make up such a product as fuselinks the ensuring of 2·0 quality level at final inspection may require an AQL of 0·4 on individual components. Since it is necessary to carry out some check on suppliers' products goods-inwards inspection should be based on an AQL of 0·4. Inspection will generally be based on batch sampling. This is particularly necessary with large quantities of small articles since 100 per cent inspection will lead to inefficiencies due to boredom. The sample sizes are based on mathematical laws of probability, which have been worked out and published in table form in such publications as UK *Defence Standard 131-A Sampling procedures and Tables for Inspection Attributes*.

Since *DEF 131-A* covers over 25 AQL figures at six different levels of inspection for normal, reduced and tightened inspection, and since the recommended sample sizes in relation to batch sizes do not follow a straight-line law, the amount of tabulated information in *DEF 131-A* appears formidable. In actual practice, however, all the data are not required in one type of production unit.

The inspection function is divided into two main stages:

(*a*)   purchased material and components
(*b*)   in-process inspection.

In addition, there is a product and process audit.

9.2 Purchased material and components

The quality of materials and components purchased from outside sources obviously has a large influence on the quality of the fuses made from them. As the purchased materials can represent a major part of the total cost of the finished fuses it follows that time and effort spent in agreeing mutually acceptable specifications is necessary to ensure that good quality fuses can be produced at economic prices. In addition the choice of most suitable suppliers has to be made.

After realistic purchasing specifications and quality standards have been agreed during initial discussions there must be periodic follow-up meetings to review the specifications and discuss any problems resulting from implementation of the quality standards.

Procedures are established to maintain effective control of all drawings and specifications in circulation, thus ensuring that all concerned are working to the latest issue. Goods-inward inspection and procedures are established by the fuse manufacturer for all material and components and detailed quality records are kept of each batch received so that inspection levels and quality standards are continually monitored and adjusted if necessary.

All rejections on all suppliers are recorded against their supplies and this record, in addition to being always available for assessment, is also to be issued as part of a periodic (e.g. monthly) report to management. The record forms the basis of vendor rating.

A further situation which arises is that the quality standards of certain parts cannot economically be checked in the fuse-manufacturer's works because of the high cost of the necessary inspection equipment or the levels of skill needed of the operators. In these circumstances the supplier is asked either to send a Certificate of Conformance with each consignment or to obtain verification from an independent laboratory to which samples are sent.

The major components of a fuselink, namely the bodies, end caps, element material, granular-filling material and, where appropriate, machined parts for the striker assembly, are checked during the goods-inward inspection as described below.

9.2.1 Bodies

A body, whether of ceramic, glass or other material, must be able to withstand the high stresses which may be developed in it when the fuselink, of which it forms part, operates in service. There should be no disruption or visual signs of distress due to these stresses, which result from the sudden rise of internal gas pressure, thermal shock and thermal expansion of the filling material.

Because of the practical difficulties of accurately simulating these conditions in relatively simple tests, two properties, namely bursting strength and porosity, have been chosen as giving an acceptable indication of the performance which will be obtained of a high-voltage ceramic fuselink body subjected to extreme conditions. A sample is fitted into a test rig which effectively seals the ends of the body. The internal pressure is then gradually raised by a hydraulic pump until the body breaks.

The internal pressure when failure occurs is noted and the value is assessed against a minimum level, set after much experience of prototype testing. The second property, porosity, is checked by placing a number of sample fragments of bodies in a pressure vessel containing fuschine dye and top pressure is applied for 24 h. After this time the bodies are examined for penetration of the dye.

Additional tests are done on low-voltage fuselink bodies, the first being a shock test during which several samples are dropped from specified heights on to a hard surface after which they are examined for resulting damage.

The ability to withstand thermal shock is checked by heating up two or three bodies to a predetermined temperature before dropping them into water or oil at a lower predetermined temperature. To be acceptable none of the samples should crack or shatter. A third test is applied to two or three samples to determine the crushing forces needed to break them, the acceptable levels again being based on experience of such testing over many years.

Dimensional checks and other examinations to study features such as surface finish are done on all types of fuselink bodies by selecting a number, dependent on the quantity in a consignment, for inspection. The length and external and internal diameters are measured using appropriate equipment such as micrometers and verniers. The most important dimension is the ground outside diameter to which the end caps are to make an interference fit.

Tubular bodies are also checked for lobing. This is done by mounting the body on two cylinders disposed to support it on two lines approximately 60° each side of its vertical centre line, as shown in Figure 9.1. The probe of a micrometer dial gauge is brought into contact with the top of the body, that is, at about 120° from each of the resting lines, and the body is then rotated through 360°.

The lobing is expressed as the total variation, from minimum to maximum of the reading of the dial gauge. Allowable variations are agreed, such as 0·0015 in on bodies up to 1 in diameter and 0·012 in on bodies between 3 and 4 in in diameter. Failure to meet the specification limits means that all bodies in the consignment have to be checked either by the supplier or the fuse manufacturer and only those found to be satisfactory may be used.

9.2.2   End caps

As explained in earlier chapters, these caps provide electrical connection between the fusible elements and the external circuit and also seal the fuselink enclosure, thus ensuring that the disturbances resulting from the operation of the fuselink are contained within it. This makes necessary the interference fits between the caps and the body, referred to above, and of course it is thus essential that the internal diameters of the end caps are within the specified limits. A number of samples are selected at random from each consignment and this critical diameter is checked using either telescopic gauges set by external micrometers or electronic bore comparators. If necessary measurements are taken at several points around a cap; other features such as overall length, hardness and surface finish are checked.

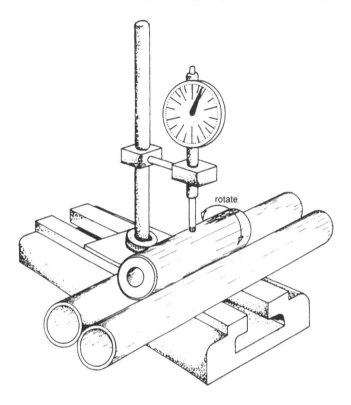

Figure 9.1 Test arrangement for checking the lobing of cylindrical fuselink bodies

Failure to meet the requirements of the agreed sampling scheme results in the caps in a consignment being returned to the supplier for all of them to be checked.

9.2.3 Element material

It is usual for fuse manufacturers to buy element material in strip or wire form and produce from it the required individual elements, because this allows flexibility in the production process.

The performance of fuselinks is heavily dependent on the element material and its diameter or thickness. Dimensions of the incoming supplies are therefore checked with comparators over standard metre lengths selected at random from each consignment. In addition, the resistances of selected metre lengths are measured, recorded and compared with specification limits previously agreed with the suppliers. This latter check tends to show that the material is suitable for use but, nevertheless, suppliers are usually required to furnish a Certificate of Conformance verifying the quality of the material in each consignment. This is done where the fuse manufacturers do not usually have suitable equipment available to measure the resistivity and determine the purity of the material.

9.2.4   *Granular filling material*

To ensure predictable operation of fuselinks under short-circuit conditions it is vitally important that the silica filling material is of high purity, a typical upper limit for impurities being 0·1 per cent. In addition, the filler-grain size must be within specified limits.

To ensure that the first condition is met, samples of each supplier's sand are taken at regular intervals and sent to an independent laboratory for full chemical analysis.

To determine whether the grain-size specification is being met, samples are taken from various levels, in a percentage (typically 3 per cent) of the bags in a consignment. These samples are mixed together and, after checking for moisture content, the material is passed through a series of sieves with aperture sizes covering the specification limits from minimum to maximum in approximately equal steps. The quantity of filling material retained on each sieve is accurately weighed and compared with the specified limits.

If the results of the sieving tests or the purity checks fail to meet the agreed specifications, the whole consignment is rejected.

9.2.5   *Machined parts for striker assemblies*

As explained earlier, certain high-voltage fuselinks are fitted with strikers which initiate the tripping of other devices. Because of the importance of this operation, the striker assemblies must be highly reliable.

Dimensional accuracy of certain mating components is essential to ensure consistent energy output. Inspection of the striker components is particularly rigorous.

9.2.6   *Components and other materials*

All other parts and materials used in manufacture are examined on receipt to a degree related to their importance in the correct functioning or life of the fuses to be produced. If a manufacturer produces more of the components in house, for example end caps, then the material for these would be inspected on arrival and the checks listed in Section 9.2.2 would be made on completion of the end caps.

9.2.7   *Calibration*

All measuring instruments are calibrated regularly against an accepted standard to ensure maintenance of accuracy, traceability and uncertainty of measurement.

9.3   In-process inspection

In-process and final inspection are conducted by the operators, 'first-off' checks of 'work in progress' are conducted for each operation followed by in-process inspection. In-process inspection is conducted on a sample basis in line with documented inspection criteria detailed in quality plans. All internal rejections are recorded on an appropriate form.

Examples of some of the more important procedures relating to high-voltage fuselinks are described in the following sections.

9.3.1   Production of fuselink elements

Most high-voltage fuselinks contain elements produced from a spool of strip metal by cutting notches or creating constrictions at regular intervals along its length. These constrictions are certainly one of the most important controlling factors in any fuselink and it is vitally important that dimensional accuracy and consistency are maintained at all times. The pitch of the notches is checked by comparison with a standard gauge length. The resistance per metre length and notch dimensions are also checked to see that they are within permitted limits, equipment such as projection microscopes being used to obtain the necessary accuracy.

9.3.2   Winding of high-voltage-fuselink elements

It was explained in Chapter 5 that high-voltage-fuselink elements are usually wound in a spiral form on supporting formers and there are often several elements in parallel. Details covering the setting-up procedures, tensioning, spacing, positioning and welding of elements are given in work instructions and supplementary documentation.

A number of visual checks and regular checks of the resistances of element assemblies during production runs are made to ensure that they are within specified limits.

'First-off' checks of each new batch and the first produced from each new reel of strip are followed by batch sample inspection. Among the features examined are the length, spacing and tension of the element, the resistance of the assembly and the integrity of the spot-welds.

Windings which do not conform to the requirements are stripped off immediately and an investigation is initiated to determine the cause of the non-conformance so that corrective action can be taken.

9.3.3   Fuselink assembly

This involves insertion and welding, soldering or brazing of the elements or element assembly into the body, filling with arc extinguishing material for cartridge fuselinks, and finally fitting of the end caps. The resistance of each fuselink is normally checked after the elements are inserted and welded to check that they have not been damaged and that the welds are satisfactory.

The filling is done in accordance with instructions provided which are designed to maintain a consistent process to cope with the many sizes and types of fuselinks and eliminate the need for large-scale checking of individual units.

Another factor which must be monitored is the degree of compaction achieved during filling. This is done by manually tapping the fuselinks and observing any fall in the level of the sand. Sand compaction is checked at the start and finish of each new batch of fuselinks and regularly during batch runs.

9.3.4   Striker assemblies

By taking the measures described to check the components and exercising tight control over assembly processes, very high statistical reliabilities are achieved.

Assembly takes place in defined stages with mandatory inspection during and after each stage. A finished batch of strikers is not accepted for fitting into fuselinks until a sample, selected at random from the batch, is tested by firing. Strikers to be tested are fitted in a pendulum rig complete with all the components normally used. The strikers are detonated electrically from a low-voltage source. The energy output of a striker under test is measured by noting the displacement of the pendulum bob against a calibrated scale. Failure to function, or an output outside the specification limits, entails rejection of the whole batch or the testing of a further much larger sample, depending on the nature of the fault.

9.3.5   In-process final inspection

Initially, this is based on batch sampling but any rejection for any of those parameters will result in 100 per cent inspection of the particular parameters of failure. Since the records are made concurrently with inspection, immediate action is possible if there is any tendency to failure. The record also forms part of the report to management.

All completed fuselinks are examined for physical defects and non-destructive tests are done.

9.3.6   Dimensional check

Because fuselinks have to fit into mountings or holders and must be capable of replacing fuselinks which have operated, their overall dimensions and those of the ferrules or tags must be within certain limits of nominal values. Clearly only the critical dimensions need to be checked during the final inspection and this is usually done using specially produced gauges.

9.3.7   X-ray examination

Radiographs of completed high-voltage fuselinks are viewed on an illuminated screen to check that the elements are not broken, twisted or damaged in any other way. The clearance between the element assembly and the body wall is checked and the low-melting-point-alloy regions are examined to see that there are no imperfections in the application of this material. The radiograph is also examined to ensure that there are no foreign bodies in the fuselink or cracks in the body walls, and the striker circuit and assembly are carefully inspected. Similar checks are done on other types of fuselinks.

9.3.8   *Resistance measurement*

The resistances of finished fuselinks are usually measured with a digital instrument with an accuracy of better than 0·5 per cent. After correcting for ambient temperature as necessary, the measured values must comply with the original design limits. For high-voltage fuselinks, the final resistance of each fuselink must agree within close limits, typically 2 per cent, with the first resistance measured near the beginning of the production process.

9.4   Other fuse parts

The other parts of fuses, such as the fuse-holders and bases, do not affect the performance as critically as the fuselinks but are equally as dangerous, and so are subject to similar component and in-process control. They are subjected to appropriate quality-assurance and inspection procedures, similar to those adopted for many other engineering products.

References

1 COCKBURN, A. C.: 'On safety fuses for electric light circuits and on the behaviour of the various metals usually employed in their construction', *J. Soc. Teleg. Eng.*, 1887, **16**, pp. 650–665

2 CLOTHIER, H. W.: 'Switchgear stages' (G. F. Laybourne and Unwin Brothers, 1933)

3 CLOTHIER, H. W.: 'The construction of high-tension central-station switchgears, with a comparison of British and foreign methods'. Paper presented to the Manchester Local Section of the IEE, 18 February, 1902, printed in Reference 2, pp. 1–19

4 KEFFORD, H. W.: 'Standardisation of fuses', *J. IEE*, 1910, **45**, Pt. 204

5 LINDMAYER, M., and SCHUBERT, M.: 'Current limitation by high temperature superconductors and by conducting polymers'. Proceedings of the 5th international conference on *Electric fuses and their applications*, University of Ilmenau, Germany, September 1995

6 BRAZZOLA, C.: 'Current limitation by circuit breaker and polymer combination'. Proceedings of the 6th international conference on *Electric fuses and their applications*, Istituto Elettrotecnico Nazionale Gailieo Ferraris, Torino, Italy, September 1999

7 ITOH, T. *et al*: 'Design considerations on the permanent power fuse for a motor control centre', *IEEE, PAG-92*, 1973, pp. 1292–1297

8 GIBSON, J. W.: 'The high-rupturing-capacity cartridge fuse, with special reference to short-circuit performance', *J. IEE,* 1941, **88**, Pt. 1, pp. 2–24

9 LEACH, J. G., NEWBERY, P. G., and WRIGHT, A.: 'Analysis of high-rupturing capacity fuselink prearcing phenomena by a finite-difference method', *Proc. IEE*, 1973, **120** (9), pp. 987–993

10 WRIGHT, A., and BEAUMONT, K. J.: 'Analysis of high-breaking-capacity fuselink arcing phenomena', *Proc. IEE*, 1976, **123** (3), pp. 252–260

11 WILKINS, R., and McEWAN, P. M.: 'A.C. short-circuit performance of uniform section fuse elements', *Proc. IEE,* 1975, **122** (3), pp. 285–293

12 McEWAN, P. M., and WILKINS, R.: 'A decoupled method for predicting time/current characteristics of H.R.C. fuses'. Proceedings of international conference on *Electrical fuses and their applications*, Liverpool Polytechnic, April 1976, pp. 33–41

13 SASU, M., and OARGO, C. H.: 'Mathematical modelling of the heat transfer phenomena in variable section fusible elements'. Proceedings of the fourth international conference on *Electric fuses and their applications*, University of Nottingham, 23–25 September 1991

14 CHEIM, L. A. V., and HOWE, A. F.: 'Calculating fuse pre-arcing times by transmission-line modelling'. Proceedings of the fourth international conference on *Electric fuses and their applications*, University of Nottingham, 23–25 September 1991

15 BEAUJEAN, D. A., NEWBERY, P. G., and JAYNE, M. G.: 'Modelling fuse elements using a CAD software package'. Proceedings of the 5th international conference on *Electric fuses and their applications*, University of Ilmenau, Germany, September 1995, pp. 133–142

16 KAWASE, Y., MIVATAKE, T., and ITO, S.: 'Heat analysis of a fuse for semiconductor devices protection using 3-D finite element method', *IEEE Trans. Magn.* 2000, **36** (4), pp. 1377–1380

17 LINDMAYER, M.: '3D simulation of fusing characteristics using the M-effect'. Proceedings of the 6th international conference on *Electric fuses and their applications*, Istituto Elettrotecnico Nazionale Gailieo Ferraris, Torino, Italy, September 1999, pp. 13–20

18 METCALF, A. W.: 'A new fuse phenomena', *BEAMA J.*, 1939, **44**, pp. 109, 151

19 BEAUJEAN, D. A., JAYNE, M. G., and NEWBERY, P. G.: 'Long-time operation of high-breaking-capacity fuses', *IEE Proc. A*, 1993, **140** (4), pp. 331–337

20 HOWE, A. F., and JORDAN, C. M.: 'Skin and proximity effects in semiconductor fuselinks'. Presented at fourth international symposium on *Switching arcs*, Lodz, Poland, 22–24 September, 1981

21 GNANALINGAM, S., and WILKINS, R.: 'Digital simulation of fuse breaking tests', *Proc. IEE*, 1980, **127** (6), pp. 434–440

22 DOLAN, W. W., and DYKE, W. P.: 'Temperature-and-field emission of electrons from metals', *Phys. Rev.*, 1954, **95**, pp. 327–332

23 SOMERVILLE, J. M.: The electric arc (Methuen, 1959), pp. 90–93

24 COBINE, J. D., and BURGER, E. E.: 'Analysis of electrode phenomena in the high current arc', *Br. J. Appl. Phys.*, 1955, **26**, pp. 895–900

25 TURNER, H. W., and TURNER, C.: 'Phenomena occurring during the extinction of arcs in fuses'. Proceedings of Lodz symposium, Poland, 1973, pp. 253–256

26 SAHA, M. N.: 'Ionization in the solar chromosphere', *Philos. Mag.*, 1920, **40**, pp. 472–488

27 SPITZER, L., and HARM, R.: 'Transport phenomena in a completely ionized gas', *Phys. Rev.*, 1953, **89**, pp. 977–981

28 BARROW, D. R., and HOWE, A. F.: 'A comparison of the computer modelling of electric fuse arcing and their real-life performance'. Proceedings of the fourth

international conference on *Electric fuses and their applications*, University of Nottingham, 23–25 September 1991

29 BARROW, D. R., and HOWE, A. F.: 'The use of optical spectroscopy in the analysis of electric fuse arcing'. Proceedings of the fourth international conference on *Electric fuses and their applications*, University of Nottingham, 23–25 September 1991

30 CHEIM, L. A. V., and HOWE, A. F.: 'Continuous evaluation of arc temperature in high breaking capacity fuses'. Proceedings of tenth international conference on *Gas discharges and their applications (GD '92)*, Swansea, 13–18 September 1992, Vol. 1, pp. 200–203

31 CHEIM, L. A. V., and HOWE, A. F.: 'The use of transmission line modelling for the solution of heat diffusion in electric fuses', *Int. J. Numer. Model., Electron. Netw. Devices Fields*, 1992, **5**, pp. 289–295

32 ROCHETTE, D., CLAIN, S., and BUSSIERE, W.: 'Mathematical model using macroscopic and microscopic scales of an electrical arc discharge through a porous medium'. Proceedings of the 7th international conference on *Electric fuses and their applications*, University of Gdansk, Poland, September 2003, pp. 188–193

33 CWIDAK, K., and LIPISKI, T.: 'Post-arc resistance in H.B.C. fuses', Proceedings of the 7th international conference on *Switching arc phenomena*, University of Lodz, Poland, September 1993, pp. 202–204

34 ERHARD, A., ROTHER, W., SDHUMANN, K., and NUTSCH, G.: 'The dielectric re-ignition of electric fuses at small over-currents', Proceedings of the 5th international conference on *Electric fuses and their applications*, University of Ilmenau, Germany, September 1995, pp. 265–272

35 MIKULECKY, H. W.: 'Current limiting fuse with full-range clearing ability', *IEEE Trans.*, 1965, paper 31 TP 65–168

36 HOWE, A. F., and NEWBERY, P. G.: 'Semiconductor fuses and their applications', *Proc. IEE*, 1980, **127** (3), pp. 155–168

37 JONES, R. A., and nine other members of an IEEE – PCIC Working Group: 'Staged tests increase awareness of arc flash hazards in electrical equipment'. IEEE Petroleum and Chemical Industry Conference Record, September 1997, pp. 313–332

38 LEE, R. H.: 'The other electrical hazard: electrical arc blast burns', *IEEE Trans. Ind. Appl.*, 1982, **1A-18** (3), pp. 246–251

39 SAPORITA, V.: 'Using current limiting fuses to reduce hazards due to electrical arc flashes'. Proceedings of the 6th international conference on *Electric fuses and their applications*, Elettrotecnico Nazionale Gailieo Ferraris, Torino, Italy, September 1999

40 CRNKO, T., and DYRNES, S.: 'Arcing fault hazards and safety suggestions for design and maintenance', *IEEE Ind. Appl. Mag.*, 2001, **7** (3), pp. 23–32

41 Safety Basics (Video) – Cooper Bussmann, www.bussmann.com (Electrical Safety)

42 STOKES, A. D., and SWEETING, D. K.: 'Electric arcing burn hazards'. Proceedings of the 7th international conference on *Electric fuses and their applications*, University of Gdansk, Poland, September 2003

43 SLOOT, T. G. J., and RILSMA, R. J.: 'Protection against fault arcs in low voltage distribution boards'. Proceedings of the 7th international conference on *Electric fuses and their applications*, University of Gdansk, Poland, September 2003

44 WILKINS, R., ALLISON, M., and LANG, M.: 'Time-domain analysis of 3-phase arc flash hazard'. Proceedings of the 7th international conference on *Electric fuses and their applications*, University of Gdansk, Poland, September 2003

45 WILKINS, R., and CHENU, J. C.: 'The contribution of current limiting fuses to power quality improvement'. Proceedings of the 7th international conference on *Electric fuses and their applications*, University of Gdansk, Poland, September 2003

46 KOJOVIC, Lj. A., HASSLER, S. P., SING, H., and WILLIAMS, Jr., C. W.: 'Current limiting fuses improve power quality'. IEEE Transmission and distribution conference and exposition, *IEEE/PES*, 2001, 1, pp. 281–286

47 KOJOVIC, Lj. A., HASSLER, S. P., LEIX, K. L., WILLIAMS, C. W., and BABER, E. E.: 'Comparative analysis of expulsion and current limiting fuse operation in distribution systems for improved power quality and protection'. *IEEE Trans. Power Delivery*, 1998, pp. 863–869

Glossary of terms

Most technical subjects tend to have a unique vocabulary. Fuses are no exception, but most fuse terms are descriptive. This Glossary of terms includes salient fuse terminology from the definitions in the following fuse Standards – *IEC 60127*, *IEC 60269, and IEC 60282*, thus providing a reference for readers.

'a' Fuselink	A current-limiting fuselink capable of breaking, under specified conditions, all currents between the lowest current indicated on its operating time–current characteristic and its rated breaking capacity
'g' Fuselink	A current-limiting fuselink capable of breaking, under specified conditions, all currents which cause melting of the fuse element up to its rated breaking capacity
Ambient air temperature (T_a)	The ambient air temperature is that of the air surrounding the fuse (at a distance of about 1 m from the fuse or its enclosure, if any)
Arc voltage	Instantaneous value of the voltage which appears across the terminals of a fuse during the arcing time
Arcing time	The interval of time between the instant of the initiation of the arc and the instant of final arc extinction
Back-up fuse (HV)	Current-limiting fuse capable of breaking, under specified conditions of use and behaviour, all currents from the rated maximum breaking current down to the rated minimum breaking current
Breaking capacity of a fuselink	A value (for AC the RMS value of the AC component) of prospective current that a fuselink is capable of breaking at a stated voltage under prescribed conditions of use and behaviour
Breaking range	Breaking range is a range of prospective currents within which the breaking capacity of a fuselink is assured

Circuit breaker A switching device capable of making, carrying and breaking currents under normal circuit conditions and also making, carrying for a specified time, and breaking currents under specified abnormal circuit conditions such as those of short circuit

Conventional fusing current (I_f) Value of current specified as that which causes operation of the fuselink within a specified time (conventional time)

Conventional non-fusing current (I_{nf}) A value of current specified as that which the fuselink is capable of carrying for a specified time (conventional time) without melting

Current-limiting fuselink A fuselink that during and by its operation in a specified current range, limits the current to a substantially lower value than the peak value of the prospective current

Cut-off current The maximum instantaneous value reached by the current during the breaking operation of a fuselink when it operates in such a manner as to prevent the current from reaching the otherwise attainable maximum

Cut-off current characteristic A curve giving the cut-off current as a function of the prospective current under stated conditions of operation

Disconnector A mechanical switching device which, in the open position, complies with the requirements specified for the isolating function

Enclosure A part providing a specified degree of protection against certain external influences and a specified degree of protection against approach to or contact with live parts and moving parts

Full-range fuse (HV) Current-limiting fuse capable of breaking under specified conditions of use and behaviour, all currents that cause melting of the fuse element(s), up to its maximum breaking current

Fuse A device that by the fusing of one or more of its specially designed and proportioned components opens the circuit in which it is inserted by breaking the current when this exceeds a given value for a sufficient time. The fuse comprises all the parts that form the complete device

Fuse base (fuse mount) The fixed part of a fuse provided with contacts, terminals and covers, where applicable

Fuse carrier The movable part of a fuse designed to carry a fuselink

Fuse combination unit	A combination of a mechanical switching device and fuses in a composite unit
Fuse contact	Two or more conductive parts designed to ensure circuit continuity between a fuselink and the corresponding fuse-holder
Fuse element	A part of a fuselink designed to melt when the fuse operates. The fuselink may comprise several fuse elements in parallel
Fuse system	Family of fuses following the same physical design principles with respect to the shape of the fuselinks, type of contact, etc.
Fuseboard	An enclosure containing busbars, with fuses, for the purpose of protecting, controlling or connecting more than one outgoing circuit fed from one or more incoming circuits (also known as a distribution fuseboard)
Fuse-holder	The combination of the fuse base with its fuse carrier
Fuselink (cartridge fuselink)	The part of a fuse including the fuse element(s), intended to be replaced after the fuse had operated
Fuses for use by authorised persons	Fuses intended to be used in installations where the fuselinks are accessible and can replaced by authorised persons only
Fuses for use by unskilled persons	(Formerly called fuses for domestic and similar applications). Fuses intended to be used in installations where the fuselinks are accessible to and can be replaced by unskilled persons
Fusing factor	The ratio, greater than unity, of the minimum fusing current to the current rating
Gate	Limiting values within which the characteristics, for example, time–current characteristics, shall be contained
General-purpose fuse (HV)	Current-limiting fuse capable of breaking, under specified conditions of use and behaviour, all currents from the rated minimum breaking current down to the current that causes melting of the fuse element(s) in 1 h or more
Homogeneous series of fuselinks	A series of fuselinks, within a given size differing from *each* other in such characteristics that for a given test, the testing of one or a reduced number of particular fuselinks of that series may be taken as representative for all the fuselinks of the series itself

I^2t (Joule integral)	Integral of the square of the current over a given time
I^2t characteristic	A curve giving I^2t values (pre-arcing I^2t and/or operating I^2t) as a function of prospective current under stated conditions of operation
Indicating device (indicator)	A device provided to indicate whether the fuse has operated
Integrally fused circuit breaker	A combination, in a single device, of a circuit breaker and fuses, one fuse being placed in series with each pole of the circuit breaker
Isolation	Function intended to cut off the supply from all or a discrete section of the installation by separating the installation or section from every source of electrical energy for reasons of safety
Minimum breaking current	Minimum value of prospective current that a fuselink is capable of breaking at a stated voltage under prescribed conditions of use and behaviour
Non-interchangeable	Limitations on shape and/or dimensions with the object of avoiding in a specific fuse base the inadvertent use of fuselinks having electrical characteristics other than those ensuring the desired degree of protection
Operating time	The sum of the pre-arcing time and the arcing time
Over-current discrimination	Co-ordination of the relevant characteristics of two or more over-current protective devices such that, on the occurrence of over-currents within stated limits, the device intended to operate within these limits does so, while the other(s) do(es) not
Overload curve of an 'a' fuselink	A curve showing the time for which an 'a' fuselink shall be able to carry the current without deterioration
Peak withstand current	(of a fuse-holder) The value of cut-off current that the fuse-holder can withstand
Power acceptance of a fuse-holder or fuse base	The maximum value of power released in a fuselink which a fuse-holder or fuse base is designed to tolerate under specified conditions
Power dissipation of a fuselink	The power released in a fuselink carrying rated current under specified conditions

Pre-arcing time	The time between the commencement of a current large enough to cause the fuse element(s) to melt and the instant when an arc is initiated
Prospective current of a circuit	With respect to a fuse the current that would flow in a circuit if a fuse situated therein were replaced by a link of negligible impedance. The prospective current is the quantity to which the breaking capacity and characteristics of the fuse are normally referred, e.g. I^2t and cut-off current characteristic
Rated current of a fuselink (I_n)	A value of current that the fuselink can carry continuously without deterioration under specified conditions
Rating	A general term employed to designate the characteristic values that together define the working conditions upon which the tests are based and for which the equipment is designed
Recovery voltage	Voltage which appears across the terminals of a fuse after the breaking of the current
Short circuit	The accidental or intentional connection, by relatively low resistance or impedance, of two or more points in a circuit
Short-circuit current	An over-current resulting from a short circuit
Short-circuit making capacity	A making capacity for which prescribed conditions include a short circuit at the terminals of the switching device
Short-circuit protector device	SCPD, a device intended to protect a circuit or parts of a circuit against short-circuit currents by interrupting them
Starter	The combination of all the switching means necessary to start and stop a motor, in combination with suitable overload protection
Striker	A mechanical device forming part of a fuselink which, when the fuse operates, releases the energy required to cause operation of other apparatus or indicators or to provide interlocking
Switch	(Mechanical) a switching device capable of making, carrying and breaking currents under normal circuit conditions which may include specified operating overload conditions and also carrying for a specified time currents under specified abnormal circuit conditions such as those of short circuit

Terminal

A conductive part of a fuse provided for electric connection to external circuits

Time–current characteristics

A curve giving the pre-arcing time or operating time as a function of the prospective current under stated conditions of operation

Transient recovery voltage (TRV)

Recovering voltage during the time in which it has a significant transient character

Utilisation category

(For a switching device or fuse) a combination of specified requirements related to the conditions in which the switching device or fuse fulfils its purpose

Index

Printed in the USA
CPSIA information can be obtained
at www.ICGtesting.com
JSHW011519221024
72172JS00008B/66